Starks' Harvesters

Starks' Harvesters

Robert S White

Old Pond Publishing

First published 2010
Copyright © Robert S White 2010

ISBN 978-1-906853-46-4

A catalogue record for this book is available from the
British Library

Published by
Old Pond Publishing Ltd
Dencora Business Centre
36 White House Road
Ipswich
IP1 5LT
United Kingdom

www.oldpond.com

Front cover: Montana 1979: the crew stage a shot of the
combines working together for photographer Charlie Norman

Cover design by Liz Whatling
Typeset by Galleon Typesetting, Ipswich
Printed and bound in Malta

Contents

Foreword by Tim Slessor   9

Introduction   13

Chapter 1   The Great Plains   17

Chapter 2   Dale Starks   23

Chapter 3   Our First Harvest   31

Chapter 4   New Combines   74

Chapter 5   Wrecked Trucks   93

Chapter 6   Long Hours   106

Chapter 7   A Canadian Connection   126

Chapter 8   Seventeen Years Later   143

Postscript   152

Author's Note

Writing about events that happened more than thirty years ago, I soon find myself submerged in a culture and language that I came to love and adopt. So I talk of truck, not lorry; wrench, not spanner; corn, not maize, and I spell plough as 'plow'. I have also tried to keep the bad language to a minimum: harvesters are apt to be coarse and profane.

These events are neither invented nor exaggerated – there would be no point. Some names and locations have been altered slightly to respect privacy.

Dedication

To my family and friends

Acknowledgements

I would like to thank the following people:

Charlie Norman for his excellent photography

Charlie, Terry & Carol Laws and Jeremy Slessor for their help in recalling these events

Tom Kirk for his recollections of harvest: he trod the path that we were to follow

Steven & Kevin Clarke and Delbert & Becky Joyner for having kept these memories alive on the Great Plains

John & Annie Chapman for suggestions and corrections to the manuscript

Tim Slessor for making all this possible

Some of the historical data has been based on material in *Custom Combining on the Great Plains, a History* by Thomas Isern.

Foreword

By Tim Slessor

ROB WHITE is an enthusiast. Indeed some might go further and say that where 'harvest' is concerned, he is almost an obsessive. Certainly, he is one of those people who, once on a course, won't bloody well let go. Now, let me tell you how I know.

More than thirty years ago I made the film *Yellow Trail from Texas* for the BBC about the American wheat harvest – probably, at that time, the biggest agricultural operation in the world. Maybe it still is. We began filming in the Texas spring; we finished nearly two thousand miles away, high up on the Canadian prairie in the first October snows. A few months later, after the film went out, I got a small flood of letters, thirty or forty of the things, from young farm lads – and a few girls – from all across Britain; they wanted to 'go and do harvest'. Please, could I put them in touch with the boss of the gang of cutters and combines that I had featured in my film?

Problem. I had to compose a standard and rather negative letter explaining that Dale Starks was a very busy man, and I doubted that he would be willing to take the risk of employing applicants he had never met and about whom he therefore could know less than nothing. Anyway, the life of a cutter was something rather less than the romantic, happy-go-lucky, sunshine existence they seemed to imagine. It was very hard work and the hours were ridiculous. And, lastly, there was the small matter of US work permits. So, please forgive me, but I'm sorry. . . .

Yes, you are ahead of me. There were two of those lads who just would not give up: Rob White and Charlie Norman. They wrote, they phoned, they wrote again; they were a regular pain. What to do? In the end, I surrendered. I told them that if they got themselves to London, I would give them lunch in a BBC canteen – how generous can one get? This, I thought would give me the chance to suss them out, and judge whether they had what it would take – to stick seven months of hot, dusty, badly paid, monotonous 18-hours-a-day-and-no-weekends, diesel-smelling grind.

Strangely, Rob and Charlie had never met before that lunch, but one could tell from the way they got on together that they were thoroughly Good Sorts. More important, they were under no illusions about what 'harvest' really involved. So I took a risk and gave them the address and details they needed. As you will learn from Rob's story, it took a year or two before they eventually made it to Texas.

Rob will tell you that his five seasons of harvest (yes, five!) are, for him, quite indelible – the best thing he has ever done. He will talk for hours, even days, about the farms he has cut, the highways travelled, the breakdowns and the accidents, the repairs and the lash-ups, the long nights and the longer days, the people met and the good friends made. It is all here in this book. He will tell you about the summer heat of Kansas and the first chills of Montana, about the mind-blowing distances and the vastness of those western skies. You will learn more about the innards and intricacies of combines than you will ever need to know. (Rob, what the hell *is* a 'shaker-shoe shaft'?)

Above all, he will tell you about the boss: Dale Starks. Not an easy man to know; a man of very few words – the quintessential sardonic westerner. Sadly, he and his wonderful wife, Margie, are gone now. But in all our memories they are still 'on harvest', still cutting. Margie, always quietly supportive, could drive a combine as well as – probably better than – any man. And Dale knew as much about working the things as anyone alive. He had started as a young man in 1948 with a wife, a bank loan and one rather clapped-out combine; by the time I met him he had nine combines and all the associated trucks and pick-ups that made up his circus.

As you may have gathered, starting with that lunch break back in 1976, Rob and I have become firm friends. He knows far more about my film than I do; he is familiar with every scene, cut and detail; he can quote long chunks of the commentary and hum along with the various songs. Amazingly, he is even now planning some kind of reunion somewhere in deepest Oklahoma with his American harvesting friends.

Another thing: when my son was in his late teens and could not make up his mind what next, I suggested that he should go off and join Rob – and Dale Starks. I remember a rather reluctant Jeremy, the proverbial callow youth, flying out from Heathrow; he returned six months later as a confident young man. He went back for a second year, from Texas to Saskatchewan. It made him. So thank you, Dale and Rob.

I hope you enjoy what Rob has written. You should. But don't ask me to enlighten you on some of the more esoteric terms and jargon that are evidently part of the harvesters' everyday patois. What *is* the difference between a Massey 860 and 850? And does it matter? How *do* you know when the distance between your cylinder bars and the concave is correct? Come again? What *do* you do when the feeder beater starts making a funny noise?

What else to say? Well, I have been lucky in that I have travelled a good deal of the world making documentaries for the BBC. But none of those films – and I would think that there have been a hundred or so – has had as long a life or has had as many echoes as 'Yellow Trail'. Who would have thought – certainly not me – that one day, more than thirty years on, it would play a part in inspiring a book. Maybe I can be forgiven for being both flattered and a little proud.

Right, Rob, get her dieseled-up and let's go!

TIM SLESSOR
London 2009

Introduction

BACK in the 1960s when I was around fourteen years old I read an article about the North American wheat harvest. It captivated and enthralled me. Driving a combine harvester across the Great Plains, two thousand miles from Texas to Canada, gripped my very being. I was desperate to do it, but having no contacts either in Britain or abroad that could help me, the chances were very slim. This was of course pre-internet days and the world seemed a much bigger place.

In January 1976 I went to work one morning and throughout the day met several people who asked me if I had seen the programme *Yellow Trail from Texas* on BBC television the previous evening. Apparently it was a documentary on the very subject, featuring a husband-and-wife team, Dale and Margie Starks, custom cutters from Oklahoma.

I had not seen it and was beside myself with grief.

So, on the strength of a programme I had not seen, I wrote to the BBC asking to be put in touch with Dale and Margie Starks.

July 1982 – The pressure was on.

The wheat in Leoti, western Kansas was ready for harvesting and we were 250 miles away in Oklahoma trying to complete our current contract. My combine was loaded behind the Jimmy Diesel truck and left mid-afternoon for Kansas. Jeremy was driving and had instructions to get as far as he possibly could in daylight, phone in on his progress and bed down for the night in the truck.

I stayed back to load the rest of the machines when we had finished that night. Then with pick-up and travel trailer I left Oklahoma some time after midnight and drove all night. I caught up with Jeremy in Dighton, Kansas, woke him up and we got the hell to Leoti. We unloaded the combine, put the header on and I started cutting wheat. I didn't quit until 11 o'clock that night. By that time I had worked forty hours without a break.

It was several days before I caught up with the boss, Dale Starks. What would he say? I had, after all, put in a superhuman effort; we had secured that job, our reputation in that area and had saved the day. Surely praise would be heaped upon me, but how would he word it?

'You made it, then,' was his only comment; but by then I should have known him well enough to expect no more. This was the Dale who had taught me about combines, harvest and, dare I say it, life itself.

So to some extent you can blame him for this book.

The North American wheat belt detailing a typical route taken by Dale Starks and his crew

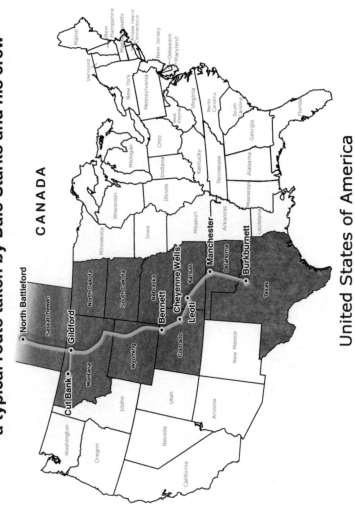

CANADA

United States of America

Chapter One

The Great Plains

B Y their geographical location and sheer scale the Great Plains of North America have lent themselves to some form of contract or 'custom' harvesting since grain crops were first grown there. Wheat is ready to cut in Texas in mid-May; crops ripen further north as summer progresses and it is October before the wheat on the prairies of Canada is finally ready to harvest. Other crops such as corn (maize), milo (grain sorghum) and soybean extend the harvest into December.

An area approximately two thousand miles long by six hundred miles wide is host to a vast quantity of grain.

In the days of the binder and threshing machine large amounts of unskilled labour – the men were known as 'Bindlestiffs' – were required for the laborious tasks of handling the sheaves from field to stack and from stack to thresher. Enterprising men bought their own thresh-ing outfit and contracted their services to other farmers. Most worked locally but a few would thresh in winter wheat areas of Oklahoma and Kansas, then load their threshing machines onto rail cars and head for the spring wheat regions further north. Custom threshing was well established by the First World War and into the 1920s.

Combine harvesters are so called because they *combine* the cutting and the threshing into one operation. Early combine harvesters were large, cumbersome machines.

17

By the 1890s machines with headers as wide as twenty feet were pulled by teams of thirty-two or more horses on the large wheat ranches of California, but it would take the effects of the First World War to see combines gain a foothold further east on the Great Plains. As wheat prices rose and labour began to get scarce a few farmers bought prairie models of twelve- to sixteen-feet cut. These were pulled by horses, mules or tractors and had engines to drive the threshing mechanism to replace earlier ground-wheel driven designs.

It was a natural progression for the owner of a threshing outfit to buy a combine to cut his own wheat – if he had any – then to hire his services to neighbouring farms. A study in 1926 by the United States Department of Agriculture found that over half the combine owners on the plains did custom work. By the 1930s some pull-type combines had rubber tyres, enabling machines to be hauled by road behind grain trucks, thus making it easier for a few enterprising operators to follow the harvest north, rather than just work locally for their neighbours.

The industry was still relatively small in 1940, but by 1942 there were enough operators for the US Bureau of Agricultural Economics to commission a study in Nebraska. They recorded the number of combines coming through the ports of entry on the southern border of that state. In 1942 they recorded 515 combines; just five years later the number had grown to 5,117.

It was the effects of the Second World War that caused the custom harvesting industry to really take off. Food was needed but manpower and resources had been diverted to the forces and to producing items such as planes, tanks and munitions. No longer were there large numbers of men available to pitch sheaves from binders. Early pull-type combines required two men, one to drive the tractor and another to ride on and operate the combine. Most pull-types had their own engine to drive the threshing mechanism and it was calculated that the average tractor/combine engine outfit used 1¼ gallons of fuel per acre harvested.

The introduction of the self-propelled combine was a big step forward: one man, one engine, using only three-quarters of a gallon of fuel per acre, and no crop run down opening up the field on the first run. In 1938 Massey-Harris introduced their first self-propelled combine, the model 20 designed by their chief engineer Tom Carroll. His improved design, the model 21, followed two years later – this was the combine that would revolutionise the custom harvesting industry.

The 21 soon became a proven product that was ripe for mass production, were it not for the restrictions on the raw materials to build it.

Joe Tucker, vice-president of sales for Massey-Harris, a man of great drive and vision, approached the War Production Board and persuaded them to allocate him enough steel to build five hundred number 21s above their normal quota for the 1944 harvest. These combines were sold to farmers and custom cutters who would pledge to cut a minimum of two thousand acres with each machine. This was to be known as the Massey-Harris Harvest Brigade and would be run as though it were a military operation.

There was no stipulation in the contract of what they would charge the farmer for cutting his wheat, but the going rate at that time was three dollars an acre. The cost of these machines was about $2,500 each. It is said that half a million bushels of grain were saved just by the advanced technology of these combines which also saved half a million gallons of fuel and made tractors and men free for other work.

The plan was to deliver the new combines by rail from the factory in Toronto, Canada to four main areas in the United States. Some went to California and the Pacific Northwest – Washington, Oregon and Idaho. About seventy machines were earmarked for southern Texas where they would begin harvesting flax and oats in April, before heading north and west to harvest wheat. The bulk of the production, about 330 machines, was to be delivered to Altus and Enid, Oklahoma and Hutchinson, Kansas: a Plains State Brigade that would be joined by the Texas Brigade and cut its way north to Canada.

Government officials and Massey-Harris dealers worked hard to bring farmers with wheat to cut into contact with custom cutters looking for work. Truckloads of parts, fuel and tyres were shipped into the area alongside Massey-Harris technicians who would ensure combines were correctly set and maintained.

It was a tremendous success: great publicity for Massey-Harris, and a favourable outcome for the War Production Board. It had been a wet, difficult harvest but the cutters had averaged 2,039 acres per machine. Massey-Harris had promised a prize for the harvester who had cut the most crop, based on dollar receipts. A $500 war bond was awarded to Wilford Phelps of Chandler, Arizona who had harvested 3,438 acres with his 14-foot-cut gasoline-powered machine.

Other manufacturers soon had their own designs of self-propelled combines into volume production. Gleaner, John Deere and International were the main contenders.

For a time there was plenty of work for everybody. In 1948 the price of wheat dropped due to over-supply. Some farmers were reluctant to hire custom cutters in these less affluent circumstances, preferring to manage with the machines they already had or to get help from neighbours. There followed a period of fluctuation while supply of combines available and wheat to cut got into some sort of balance. Some operators drawn by the profits of the good times fell away when times were hard. But by now there were a substantial number of professional operators who were economically and emotionally tied to their jobs, just as the farmers were tied to their land. Did a farmer not plant his crop even though prices at harvest were predicted to be low? Some cutters made the run, come what may.

The 1960s saw more stability in the industry. There was always the risk of crops being lost to hail or drought, but cutters modified their routes to suit conditions and demand.

From 1973 to 1976 high yields combined with high prices ensured good profits for both cutters and farmers. Wheat was being cut for about eight dollars an acre at this time and established operators invested in more sophisticated and expensive machinery. When the price of wheat dropped in the middle of 1976 some cutters found they had borrowed too much money to finance machinery, but the weather in 1977 again played its part and heavy rains in Texas, Oklahoma and Kansas kept cutting rates about the same as the previous year.

An outfit might consist of one combine and one grain truck or up to ten or more combines, each with a grain truck. It was a male-dominated industry in the early days, but wives began to accompany their husbands on harvest to drive truck and cook the meals. Children too came along during their long summer school vacation. Lasting friendships would become established between harvesters and their farming customers. Wives would meet up and children play together which, as well as doing a good job, ensured continuity of work in future years.

Living accommodation improved as families came along. In the early days men would sleep in tents, barns or granaries. Travel trailers became popular, the boss and his family living in relative comfort with facilities for the wives to produce those all-important meals for a hard-working crew. The crew's accommodation would be more basic; a trailer house with bunk beds might have looked grim inside, but it was a welcome sanctuary after a long, hard, hot day on the combine. Old school buses were another favourite. Bought second-hand they were cheap and easy to convert to mobile bunkhouse accommodation, their yellow hue a sharp contrast with the combines and trucks they accompanied.

Chapter Two

Dale Starks

DALE STARKS was born on 17 January 1924 to Wallace and Ethel May Starks. The Starks farmed in northern Oklahoma, close to the homestead that would eventually become Dale's farm. Dale was one of seven children: three girls and four boys.

In 1943 William Dale Starks, aged nineteen, married Margie Fearn Witters aged sixteen, daughter of a construction worker. The Witters lived a few miles south of the farm. Margie had one brother, their mother having died when Margie was seven years old. The newlyweds lived in half of Dale's grandmother's house and they began farming the 160 acres of land which, like the house, was rented.

In 1948, with the aid of a bank loan, Dale began custom harvesting. The previous winter Dale had bought and worked on his first combine. It was a used pull-type Gleaner, 12-foot cut with a 6-cylinder Hercules engine. This he pulled with a Case D tractor. When moving between jobs the tractor was loaded onto the grain truck with the combine towed behind the truck. The header was loaded onto a trailer and hitched behind the combine. He employed one man to drive the grain truck, and that first year they harvested in Hobart, Oklahoma; Manchester, Oklahoma, and Ellinwood, Kansas.

1949 saw the same outfit harvesting in Frederick, Oklahoma; Manchester, Oklahoma; Harper, Kansas; Scott City, Kansas, and Sterling, Colorado. In Sterling, Dale was cutting sunflower-infested wheat alongside another cutter who was using a self-propelled Massey-Harris 21 combine. Dale's Gleaner was having a lot of trouble harvesting the crop, but he was very impressed with the Massey's engineering and performance in those adverse conditions. Dale hated his Gleaner with a passion. He claimed to have had only two happy days in his life, the first being when he sold that combine. When pressed for information on the second, he said it was the day he sold the tractor that pulled the 'son-of-a-bitch'!

The Gleaner had a slatted conveyor to take the straw away from the concave area. In heavy crop conditions the slats would break and it was an awkward and time-consuming operation to replace them. The combine had only one slip clutch to drive the whole mechanism. If it was backed off enough to avoid breaking slats it wouldn't run the machine, if it was tightened up enough to run the machine it would break slats. Dale was not impressed, but he must have made some money that year because he bought a new pick-up.

In 1950 Dale bought a brand-new Massey-Harris 27 combine, successor to the 21. He had a Ford truck. The combine, with 14-foot header still attached, was driven onto the truck for moving between jobs. With the header over the cab roof, the truck must have been top heavy and presented a fearful sight to oncoming motorists. That year they cut in Hobart, Oklahoma; Manchester, Oklahoma; Scott City, Kansas, and in Sidney, Nebraska for a man by the name of Harry Sparks, a job Dale was to do for the next twenty-seven years.

Then for the first time they made the epic journey to

northern Montana, to what is known as the Hi-Line, so-called because it is the most northerly highway running east–west. He cut near the town of Kremlin, west of Havre. Dale was hired to harvest part of 1,200 acres of wheat belonging to Lenny Milner. Milner did not actually farm himself but had a man and his wife living on the farm who could be relied on to run the place unsupervised. That couple were Wade and Edna Reese, eventually to become life-long friends of the Starks.

Wade, who had migrated from Missouri at the age of seventeen, acquired some land of his own over the years and wooed and married Edna. He worked for Milner for twenty years, running the place as though it were his own. Indeed the plan was to buy Milner out eventually, but fate intervened. Milner's wife had a stroke, rendering her bedridden for twenty-six years and unable to speak or feed herself.

This obviously had an emotional and financial effect on Lenny. A local realtor and farmer who already owned thousands of acres talked Milner into selling him the farm, thus depriving Wade and Edna of their chance. They left their beloved farm, north-west of Kremlin, and bought a house in nearby Gildford. Wade got a job with the county driving a road grader for the next seven years. Wade became a great help to Dale in securing work for him in that area.

1951 saw Margie's first harvest. Until then she had stayed at home to run the farm. It was Margie's job to mouldboard plow and when their son Larry was quite young – born, I think, in 1950 – he was pulled alongside the tractor in a wagon as she worked the land. This arrangement was necessary, as there was not enough room for Larry to be on the tractor. Larry was named after Dale's brother, Lawrence, who was killed in a car accident when he was seventeen. The driver of the car was drunk, they ran off the road, overturned in a ditch and Lawrence was drowned.

Dale's farm came up for auction that year. At that time he was cutting in Sidney, Nebraska for Harry Sparks. He cut all day, hired someone to drive him the 450 miles back to Manchester, Oklahoma overnight while he slept, bought the farm and was driven straight back to resume cutting. He paid $29,500 for the house and 160 acres.

They did not make it to Kremlin that year as they got rained out further south. By now he was running two machines.

In 1952 Dale traded for a new Massey-Harris 27. It was either this year or 1953 that he worked with a man by the name of Ben Steinmetz. Ben and his wife Sadie owned their own 27 and cut alongside Dale for two or three years.

By 1954 Dale was running four machines and this was the first year he cut for the Goetze brothers of Iowa Park, Texas. He would cut for the Goetzes for the next thirty-seven years. Since Margie had started to come along they had lived in trailers, but after a day on the combines they were still taking their bath in the ponds used for watering cattle. Dale said that in the early days he slept in the granaries along with his truck driver – if the mosquitoes got too bad they would cover their face and arms with diesel oil.

By 1956 Dale was running six machines. The Massey-Harris 90 was now in production, costing $5,000 each. Cutting and hauling in Montana brought in five dollars an acre. Good money was made out of custom harvesting in those years.

Dale had become well established in Montana. Along with Milner's wheat he was also cutting 1,300 acres at Kremlin for Vernie Hill. Vernie had arrived in Montana in 1911, full of hopes and with just fifty dollars in his pocket. At the age of twenty-three he left North Dakota to seek a better life. Those early years were tough. He built a little homestead shack on his 'free' 160 acres and that first winter he trapped badgers to survive.

By working for other people in the winter and doing some taxidermy he survived where others failed. Being shrewd and smart, Vernie bought up surrounding homesteads as they became vacant, eventually farming about 3,000 acres.

His daughter, Dorothy, married Howard Bailey who gradually took over the running of the farm and in the film 'Yellow Trail' talks of Dale being 'the cream of the cutters'. Howard's daughter Jeanne always looked forward to Dale and his crew coming to harvest their wheat. She said their farm would come alive as combines and crew pulled into their yard.

The Massey-Harris 92 was introduced to the fleet in 1958, followed by the Massey-Harris Super 92 in 1960. Massey Ferguson Super 92s would follow. Six machines were still being run by Dale at this time. In 1964 two 410 combines were bought, as well as a John Deere model 95. The Deere was found to be short of horse-power and traded after two seasons.

It would seem that the 410s were gasoline-powered, as Dale talked of his first diesel combines being the 510s, arriving in 1966. These machines had 20-foot headers. It was on the 510s that Dale had his first cabs. It seemed like a good idea, but the Texas heat in a cab with no air conditioning proved too much, so they removed them when they got back to Manchester. Later that season, in the fall, Dale sent a truck back south to Oklahoma to collect the cabs, which proved very welcome as temperatures plummeted.

For transport purposes the 510s were still compact enough to be driven onto the grain trucks after the trucks' grain sides had been removed. The headers would be loaded onto a trailer, which would be hitched to the back of the truck after the combine was loaded.

Dale's heyday was while he was running 510s. His reputation at that time was second to none. Farmers would ignore other cutters sat in town looking for work, and wait for Dale Starks to arrive.

Early in his career he met Tom Carroll, designer of the Massey-Harris 21 combine, and held him in high esteem. Later on in the '80s I was to meet Frank Vieira and Kerry English of the Massey Ferguson Harvest Brigade, both of whom Dale held in great respect, and they in turn respected Dale. If you visited the Massey Ferguson Brigade trailer and said you were Starks' crew you were listened to.

Wade told me that Massey had offered Dale a job trouble-shooting for them: to visit new machines, solve problems, and suggest design improvements. He was offered a huge salary plus expenses but turned them down flat.

It seems likely that Dale bought his first 760 and 750 combines in 1973, as he remembered buying one of each to try out. Ten 760 combines were built in 1971 and the first main batch of five hundred 760s was produced for the 1972 harvest. As the 750 was not built until the following year, it is doubtful that Dale had one of these first five hundred machines.

As these combines were too big to be driven onto the grain trucks, a new system was devised to transport the 700 series machines. The end gate (tail board) of the truck was first removed. The header was taken off the combine and a home-made crane jib called a boom was fitted to the feeder-house of the combine. A chain was bolted to the header and the combine lifted the header high enough to allow the truck to reverse underneath it. Truck boxes were 20 feet long, headers 24 feet, so a 4-foot length of header hung out of the back of the truck.

Then a home-made trailer or 'lowboy' was hitched to the truck and the combine driven on. This meant that one man could haul his combine and operate alone if necessary. It was easier to manage the harvesting operation if all machines could be kept together, but often one machine might go ahead to secure a job that was ready or one stay back to finish while the rest went ahead, or maybe the crew would be split in half. By the time I worked for Dale in 1979 he was running all 760s and 750s, his last two 510s having been sold at the end of the 1975 harvest.

The lowboys had eight wheels on four short axles and if a tyre blew, the axle would be angled enough for the other wheel to chafe on the trailer frame. If you did not have a spare with you, you had to jack the trailer up and chain the axle up to the trailer frame. Someone suggested chaining all the axles up, removing all the wheels, thereby solving all future tyre trouble!

It was best if you could move on a wet day. Not only were you not wasting cutting time, tyres ran cooler too. Also, in those days most of the trucks ran on gasoline which, if you were pulling heavy loads up steep hills in very hot weather, could vaporise in the fuel pump before it got to the carburettor. The only cure for this was either to let the engine cool down or to pour liquid onto the fuel pump. In that heat the contents of your water jug – usually Country Time lemonade – were far too precious to use. So if you ever saw a harvester relieving himself on his truck engine, you now know why he was doing it!

Road regulations at that time did not require the lowboys to have brakes of any kind, so going down steep hills was undertaken with care. Some of the trucks were getting old, their roadworthiness leaving something to be desired. In Wyoming we hauled 10-tonne machines up and down some pretty rugged country, and in Montana the Missouri Breaks north of Grassrange were a test of man and machine.

One year I drove the '68 Chevrolet drag axle to Montana pulling a combine, and by checking fuel tank capacity with miles travelled I worked out the gas consumption. On a hundred-mile stretch, in the steepest terrain, in very hot weather, and with an engine desperately out of tune, it was doing four miles to the US gallon.

In 1974 Dale bought a cattle ranch west of Springview, Nebraska, some five hundred miles from Manchester. It consisted of two full sections of land (1,280 acres), and later a further 970 acres at Wood, South Dakota were added. There was no house on the land; Dale and Margie lived in the new 26-foot Red Dale trailer to start with and bought the new larger Mayflower trailer in 1976.

Chapter Three

Our First Harvest

AN August 1972 article 'North with the Wheat
Cutters' by Noel Grove in the *National Geographic*
magazine caught the attention of the BBC television
'World About Us' series editor Anthony Isaacs. He
asked producer Tim Slessor to investigate the possibility
of making a film about the North American wheat
harvest. Tim was fascinated by the history and culture of
the High Plains, having in the mid-'60s interrupted his
television career for a year to teach in Nebraska.

In the summer of 1975, when Dale was running nine
machines, Tim made three trips to the States and with
different film crews shot *Yellow Trail from Texas*. The
film captured the imagination of a lot of viewers, many
requesting to be put in touch with Dale Starks with a
view to working on harvest. Understandably they all
received a standard reply saying Mr Starks was a busy
man, combine driving being a skilled operation, and that
harvesting could be a pretty tough life. This and following letters put all but two people off. Charlie Norman
and I would not give in.

We had said we would work for nothing and sleep in
a truck if necessary. Tim decided to put us in touch with
each other, arranged to meet us at the BBC, showed us
the film and took us to lunch. For me this was the first
sighting of the film – I was enthralled.

It took three years of letter writing and phone calls to make it actually happen. Dale was keen to hire us, but US work permits and immigration were the problem.

Charlie and I were from very different backgrounds. Although not from a farming family, Charlie had helped on several local farms during school holidays and had a passion for the life, the machinery and especially the combines. On leaving school he had started work on a farm that was to be a year's practical experience before going to agricultural college. However, he got side-tracked, went to France for a few months, then returned to England to join Lloyd's of London as a junior clerk. By 1979 he had progressed to being a broker, but his heart was in agriculture, not the City.

I was brought up on a small rented farm in Derbyshire but in 1969 my father's health and our financial position saw us leave the farm. I left school at fifteen and got a job on a large arable farm in Norfolk. After five years I had saved enough money to buy two second-hand tractors and a fertiliser spreader and was able to start a modest contracting business. The work was seasonal, giving me the opportunity to do other work in the summer.

At last, in May 1979 Charlie and I flew to New York and caught Greyhound Bus 8243 to travel the 1,500 miles to Wichita Falls, Texas where we had arranged to meet up with Dale. New York was as far removed from harvest as you would expect it to be. As we stood in line at customs, an officer angrily accused a lady of jumping the line. He picked up her bag and threw it across the room. I wondered what we had come to. Like all other cities, I was pleased to be out of it.

The Greyhound Bus company is a very professional outfit. The bus was clean and the staff polite and helpful. 'Ridin the Dog' as it is known, was a new experience to

us and an economical way to travel. Our bus journey
took us through varied landscapes, but we were nearly at
our Texas destination before we saw the first combines.
My heart leapt. Combines on lowboys hauled by trucks
were to be seen going down the highway in twos and
threes. This was a world away from New York, this was
it – we had finally made it.

From Wichita Falls bus station we called Dale to dis-
cover that he had not yet left his base in Oklahoma. He
suggested we check into a motel for a few days until he
came south but, after putting the phone down, Charlie
and I decided this was not a good idea. We caught a bus
to Nash, some thirty miles south of Manchester, called
Dale to announce our arrival, and were picked up by
John Costello, one of the crew members.

The few hours spent in the small town of Nash gave
us a taste of the friendliness we would encounter in these
small rural communities. When we had asked to use the
phone at the local gas station we found the proprietor
very friendly and helpful.

As customers called in for gas he would introduce us
to them. Obviously it was unusual for two very English
speaking people to be in town, and word soon spread.
The local Pastor came to see us, and a farmer told us all
about the 'Army Worms' that were invading his wheat.

When we arrived at the farm we met Dale. I felt at
ease with him immediately. It was not difficult to
convey our enthusiasm for the task ahead, and although
we were a continent apart in culture I felt that all was
going to be okay.

Margie was conspicuous by her absence: she was five
hundred miles away on the ranch in Nebraska. This was
the first year she did not go on harvest since 1951, and
meant that the domestic side of the harvesting operation

was going to be changed dramatically. Dale had hired the Costello family that year. Don and his son, John, were to be drivers, Jerry was to be the cook.

Having spent twenty-four hours in Wichita Falls bus station with little sleep, we were very tired on arrival at Manchester. After a meal and a long talk with Dale he showed us the bunker. It was a trailer with eight bunk beds where the single guys would sleep. Apparently, it had just been brought back from Kansas where it would have been used on the corn harvest the previous fall. The unsecured table and chairs were scattered all over the floor and it was in a hell of a mess, but it looked pretty damn good to Charlie and me who had not stretched full length and slept properly for two days. The rest of the single guys were sleeping in an old house across the fields, so we had the place to ourselves that first night.

We soon settled in and made good friends with the crew, including Terry Laws, a twenty-year-old farm boy from Missouri who had seen 'Yellow Trail' on US television and been inspired to do a harvest run. Terry contacted Massey Ferguson in Kansas City who gave him Dale's number. When he spoke to Margie on the phone at the ranch, Dale had just left for Manchester. Margie told Dale to hire him on, and Terry was offered the job. Terry was to work for Dale for the next four years and become his right-hand man.

We spent our time preparing combines and trucks for the onslaught. Only one crew member, Danny, alias 'Slim', had done the harvest run before, but not with Dale. Slim was a likeable character, but basically a drifting alcoholic in his late thirties. We were not to know it, but most of that crew would not make it to the end of the season.

Each January Dale would place advertisements in newspapers for crew members. One or two might be farm boys used to operating equipment, but most would have little or no experience. Very few returned to do a second harvest: it was a tough life, the hours were long and the pay was poor.

Dale ran five Massey Ferguson combines that year, a guy called Terry Pfeiffer running two machines along with us. Dale had three 1977 760s, a 1976 750 and one brand-new 750. Terry had one unused 1978 760, and another 1978 760 that had completed one harvest.

Trucks that year were: '77 Chevrolet tandem; '74 Chevrolet tandem; '73 GMC Diesel; '70 Chevrolet twin screw; '68 Chevrolet drag axle; '67 Chevrolet 4-wheel bobtail, and a '66 Chevrolet parts truck. There were two Chevrolet Scottsdale pick-ups, one of which Dale drove. It had toolboxes in the back, a mobile phone and two-way radio installed, the top of the dashboard serving as his office. The other was Margie's, a 1975 model with a 7.25 litre engine. Although driven by any crew member, it was always referred to as 'Margie's pick-up'.

When travelling between jobs, the grain trucks hauled a combine each, the parts truck hauled the bunker, and pick-ups pulled the two smaller travel trailers, the Red Dale and the Terry Trailer.

Don Costello, an ex-garage mechanic with Chevrolet experience, was rebuilding the engine in the '70 twin-screw truck. Charlie and I were pleased to be doing simple work, sometimes things as mundane as cleaning down a truck bed ready for re-spraying. Of course, we regularly sat in the combine cabs to familiarise ourselves with the controls.

Dale would not say who were to be combine drivers,

most people's first choice, and who would drive truck. He was far too astute to commit himself at that point, but he was assessing the crew all the time, and was a pretty good judge of people. We lost two crew members before we even left Manchester. Cliff, single in his twenties, quit after a few days. A married couple whose names I can't recall decided to hitch up their trailer, retrieve some catfish that Jerry was storing in her freezer and head for pastures new.

After about a week we loaded the machines and headed for Burkburnett, Texas, some 250 miles to the south, down by the Red River where the wheat was close to being ready for harvesting. I was glad we had not waited in Wichita Falls for them to come south, for we had seen Dale's base in Manchester, got to know everybody and had built up some confidence at the start of our dream adventure.

There were enough drivers for all the trucks and pick-ups to enable Charlie and I to ride shotgun with other people. Charlie rode with Slim in the GMC Diesel pulling combine. I rode with a guy called Gary Schneider in the '77 Chevrolet tandem, also pulling combine. I for one was pleased not to have to drive anything on this first trip.

Our journey south was not without incident. The '70 twin screw's rebuilt engine was making noises after just a few miles. The convoy stopped – we were still on dirt roads – and Gary and I hopped out to see what the problem was. We had left our truck engine ticking over, and when everybody was ready to continue we jumped back into our cab to find the temperature gauge on red. First lesson learnt. Do not leave a gas truck's engine ticking over in hot weather; without forward motion it will get hot – switch it off!

Somebody had a flat in a lowboy further down the road and by the time we hit Ringwood, the '70 twin screw required a new engine.

Dale decided we should split. He would stay back to arrange for a new engine to be fitted, Slim knew the way and would lead the convoy. I remember Dale going over the route with him: Watonga, Hinton, Anadarko, Apache and then south on 44 to Burkburnett. These unfamiliar place names just added to the thrilling anticipation of our adventure. What would happen next I wondered.

As darkness fell, we stopped in Okeene for the night. Our combines, even without headers attached, were over-width, and the law required we pull over one hour before sunset. It was rarely a problem to park the convoy, however big it was. There were usually vacant lots in town, the Massey Ferguson dealer was a likely place, and the roads were so wide in town you could very often just pull over on the side.

Back in Ringwood, Don needed to fit a boom on the combine he was pulling so he could utilise the combine to lift the old truck engine out and the new engine in. Not all combines were fitted with a boom; we had three in total, and maybe only had one with us that trip. Consequently, that evening Terry and Charlie ran the one we had with us in Okeene back to Ringwood in Margie's pick-up.

We slept in the bunker that night, not of course hooked up to either water or electricity. Our chuck wagon, Jerry and Don's trailer, was back in Ringwood, so we ate in town. The next morning, while we were waiting on Dale's okay to continue south, somebody organised the hire of a motel room for one hour. There was a production line of toilet – wash basin – shower,

37

and we all emerged feeling and smelling somewhat fresher!

The next day we pulled into Burkburnett where Dale and company caught up with us. As it was wet at the farm, some eight miles away, we parked on the edge of town for, I think, two days, awaiting the dirt roads to dry out to enable us to get to our first contract.

With the anticipation of what was in front of us all the crew were in high spirits, something that nearly got us into trouble. About eight of us were eating in Sargii's restaurant and there was the usual banter with the waitress who happened to be the proprietor's daughter. We placed our order and waited, laughing and joking among ourselves. Boisterous yes, but not rowdy.

We waited quite a long time, noticing that some other customers got served before us although we had come into the restaurant before them. Gary jokingly challenged the waitress, who said that one of our crew had ordered chicken and that would take longer.

'Well, fuck the chicken, the rest of us need to eat now,' was Gary's reply. It was said in jest but things nearly turned ugly. We did retrieve the situation, but a big, stocky guy came from behind the counter to sit on a barstool, drink his coffee, smoke his cigarette and scowl in our direction. I think this was Sargii, he certainly looked like an ex GI. Nothing was said, we didn't ask!

It took a few more days before the wheat was ready to cut. Not that we were idle in that time. Remember that we had only one new combine that year, so there was plenty of adjusting to do on combines and trucks.

At last the wheat was ready to try, but there was still no allocation of men to machines.

Dale told Slim to take one of the 760s and cut a sample in the field right next to the farmyard. What a joyous sight to see – a real live '60 cutting ripe Texas wheat on a very hot afternoon. This is what we had dreamed of, this is what we had come to do. The harvested wheat sample was checked for moisture content and that it was being threshed cleanly from the straw. It was declared okay and a truck was made available for Slim to dump the grain into.

Dale looked at me and asked if the 760 behind me was ready for work. I confirmed it was, and Dale said, 'Think you can handle it?'

I said yes I could and without delay clambered into the cab. I had expected Dale to ride with me a round, but he made no sign of joining in, so I carried on. I had felt at home the moment I first sat in one of those combines, everything coming naturally to hand. So now we had two machines on that 220-acre field known as 'The Snow Place' and gradually, one by one, men were put to machines until we were all cutting.

Oscar and Ernest Goetze (pronounced Gettsy) were very friendly and a joy to work for. They were the hard-working, hands-on type of farmers. Our agricultural backgrounds seemed to bridge any culture gap. They both had a great sense of humour. Oscar said, 'When a farmer goes to town on market day he is like a stray dog. If he stands still he gets screwed, and if he turns to run away he gets his ass kicked!'

★   ★   ★

We soon settled into a routine of long hours combining, and not much sleep.

In Texas there was no summer fallow, and no strip farming of the sort we would encounter further north. The whole farm, scattered over about ten miles, was seeded to wheat each and every year. Field size varied from 5 acres, invariably called a 'peanut patch', to 640 acres. Some were of irregular shape, interspersed with areas of scrub and mesquit bushes. The mesquit (pronounced moskeet) had large thorns on them and if we had to go near them to access a field, great care had to be taken to avoid a flat tyre.

Beef cattle fed on some wheat fields in the early stages of growth, then moved onto pasture to let the wheat grow on and mature. I use the word 'pasture' with caution. Sun-scorched scrub would be a more appropriate term. Oil wells with 'nodding donkeys' were a feature of the area and, in two locations, a capped oil pipe stuck out of the ground just below the top of the wheat, with the crop seeded very close to it. Imagine what that would do to a 24-foot header if you hit it. They were in fields of 100–200 acres and very hard to find.

The Goetzes originated from Germany where their grandfather Ernst had kept a tavern. He was concerned for his family – three boys and a girl at that time – as their country always seemed to be involved in wars, and he had heard of the opportunities available to pioneers on the plains of America.

In 1886 he sold all he had and the whole family sailed to New York, took a boat to Houston, Texas, then travelled by train to Denton, just north of Dallas. There he bought a team of two horses and a wagon and ventured another 175 miles in search of a place to settle. They finally stayed near Wichita Falls, north-west of Iowa Park and south-west of Burkburnett. It seemed a good place and the fact that trees grew there seemed to help them decide that was where they would be.

They filed on forty acres for 27 cents per acre, and dug a hole in the ground that became their home for the first year. Another child was born in the dugout but did not survive. That first year they had eleven months of drought and their crops and garden failed. A neighbour from the Burnett ranch came by one day to see how they were getting on and, seeing their plight, said nothing but returned home and instructed a cowboy to drive an old cow to their dugout as a present. They butchered and ate the cow, probably saving their lives. One more girl was born and survived.

About a year later the eldest child, Max, father of Oscar and Ernest, who was by that time fourteen years old and his brother Paul, a year younger, took the team of horses to Louisiana to work in the rice fields for a year. All the money they earned was sent home to Texas to help keep the family alive.

Over the years more land was bought and soon they were farming three hundred acres. They raised oats, wheat, cotton, a vegetable garden for the house and ran a few cattle.

Max bought 'The Home Place', some two miles south of the dugout, moved a wooden barn onto it to live in and, in 1904, married Margaret Streit, from Vernon and of Swiss/German origin. A stable was built for the horses and the original barn dwelling of two rooms was added onto over the years as the family grew. Six boys and three girls were born in the following order: Fritz, Max, Rosa, Carl, Emma, Paul, Margaret, Oscar, and Ernest.

By the time Max died in 1946 he owned one section – 640 acres – of land. Margaret died in 1956. Oscar married and moved to a homestead about five miles from The Home Place in 1954, bought it in 1956, and a new house was installed on site in 1975. By the time we were cutting for the Goetzes they were growing about 2,600 acres of wheat. Fritz, Paul and Ernest never

41

married, and lived in the original wooden homestead house until they passed away.

We had parked the bunker and the other trailers in Oscar's yard. Dale slept in the Red Dale trailer, we ate breakfast in Jerry's trailer. All were hooked up to water, electricity and bottled gas. The bunker had a shower fitted, which might have to accommodate ten or so harvesters at the end of the day. It was very spartan inside with no air conditioning, comfortable chairs or TV. There were eight bunks – four units, two tiers high – of basic metal construction with a latticework of metal wires, criss-crossed to form a base.

On this was a foam mattress about six inches thick, roughly cut to size and not benefiting from a cover of any kind. Most crew members had brought their own sleeping bag, but there were a few blankets around for those who had not. I don't think those blankets saw a washing machine from one year to the next. Pillows were unheard of; I used a rolled up coat. The general state of the bunker reflected the fact that eight single guys slept in there. Cleanliness and tidiness are not always top of the agenda for a harvest crew.

Oscar's old homestead house was still in existence, though no longer lived in, together with its original out-house. These shacks worked okay; there were enough gaps in the wooden boarding to keep the smell at a bearable level. One year a young and innocent crew member asked Dale about 'bathroom facilities', Dale nodded in the direction of the shack. 'Oh I couldn't possibly use that,' said the young hand. 'Well,' replied Dale, 'it's never refused anything I've offered it.'

The bunker did have a toilet – open plan – but sewer connections were not always available.

Dale's first job of the day was to open the bunker

door and rouse his crew. It was usually good humoured, suggesting we might get calluses on our asses if we stayed in bed any longer. Another favourite was to say that we were getting our days and nights mixed, it was daylight and we should be up. 'Does this mean then that we quit at sunset tonight, rather than work on till past midnight?' someone enquired.

I loved working after dark; the world took on a new dimension. Watching that wheat fold under the auger illuminated by six powerful work-lights was magic. The silhouette of other combines and trucks at their work was majestic. Combines ran better in the cool of the night, though the temperature was still in the stifling eighties.

Sometimes I would slide my side window open and listen to that British-built Perkins engine roaring away in the night; a comforting, dependable sound. In 1958 Massey Ferguson had bought the Perkins diesel engine company in Peterborough for $12.5 million. I felt rather proud that this combine manufactured in Canada relied on British engineering to power it.

Jerry's breakfasts, eaten around seven o'clock, usually consisted of bacon, eggs, sausage, toast etc., and sometimes 'biscuits and gravy', a traditional American meal which had no connection at all with its English counterpart. There was lots of coffee, black and strong, with hot tea made especially for the likes of me who dislikes coffee and could not bear the stuff that they drank. The food was excellent.

After breakfast she would present everyone with a 'sacked lunch', usually ham sandwich or similar, a piece of fruit and maybe a few cookies, all contained in a brown paper bag.

Everybody was issued with a water jug. This was a

plastic insulated container, one US gallon in capacity, usually made by Thermos. Before we left for the fields we would make sure this was filled with, at the very least, iced water, or preferably Kool-Aid, or Country Time lemonade, a refreshing non-fizzy beverage. Dale would point out the futility of drinking Cola and the like, saying that all that sugar would not quench your thirst, and he was right.

Daytime temperatures were already into the high nineties (high thirties centigrade) and we were told it would get hotter. Water jugs were important. Running a combine with a broken air-conditioning system, I have drunk two jugs a day, and had little need to wet the combine tyres during that time.

Lunches were eaten 'on the go'. Around 7 pm Jerry would bring supper out to the field, a hot meal of meat, vegetables etc. We would stop our machines just long enough to eat it, though leave the engine idling to let it cool down and also to keep the air conditioning operational. After fifteen minutes or so Dale would suggest we were 'burning daylight' and we would clamber back into our cabs.

Our working day would begin after breakfast in Oscar's yard. The parts truck or 'Caddy' as it was known, because it was our Cadillac, had a large fuel tank on board which was filled from the huge diesel tank in Oscar's yard. This was then driven out to the fields and the combine fuel tanks were replenished. Fortunately the Caddy had an electric fuel pump installed: the days of hand pumping were long gone. For the distances involved it was not practical to bring the combines back to the yard at the end of each day for servicing and refuelling.

Loaded trucks from the previous night were driven

into town to be dumped at the elevator – the huge concrete silo storage facility – while we combine drivers serviced our machines. Combines were usually parked at night, or in the early hours, in a straight line facing west so that the fierce morning sun was not on the cabs. Headers were left raised high in the air – no health and safety at work rules – to stop rattlesnakes or skunks crawling into the machines. Being cold blooded, rattlesnakes would seek the residual heat of a machine overnight as the air temperature dropped and might prove a nasty surprise when you greased the combine the next morning. If you accidentally ran a skunk through the mechanism it would stink for days afterwards.

We would try to park on bare ground if possible, in case a fire was caused by lightning strikes. Dry storms were quite common in this area. With such hot days and vast expanses of tinder-dry wheat and stubble, a flash of lightning could set off a fire of monumental proportions.

Greasing and general servicing were undertaken with plenty of jolly social banter. 'Could you spare a grease cartridge?' or 'Should your reel bats have that much damage to them?' together with extolling the virtues of having one's exhaust stack facing forward rather than upright, or, horror of horrors, backwards! This was a good time to interact with crew members before we were isolated from each other for the rest of the day. Though we would be cutting in the same field and have two-way radio contact, alone in your cab you had plenty of time to collect your thoughts, which was fine by me.

We carried a good supply of spares in the compartments of the Caddy. Sickle (knife) sections were checked daily. There are ninety-six of them on a 24-foot header. Texas is rocky and we broke the occasional one, but as the season progressed we simply wore them out. A wet day would give you a chance to get to grips with them, but I would try and replace say five each morning if we had been running for a long period of settled weather.

Engine oil and water levels were checked daily and replenished if necessary. Chain and belt tensions were also checked daily and adjusted as required. Oil levels in gearboxes and hydraulic tank were monitored perhaps weekly or whenever you thought there might be a problem. In those dusty conditions engine and cab air filters needed blowing clean, sometimes daily. Two of our trucks had a flexible airline hose attached to their airbrake tank for this very purpose. Some custom harvesters had a portable petrol-engine-driven air compressor on their service truck. As this was open to the elements rain would rust the pressure relief valve – and I heard more than one story of a compressor exploding, sometimes with fatalities.

I adopted a routine for greasing, starting at the header on the left-hand side and working round the machine to finish at the opposite end of the header. It soon became easy to spot potential trouble. If a part was working loose and had moved slightly it looked wrong and was soon detected. Main bearings got ten strokes of the grease gun every six to eight hours. The sickle bearing which did a lot of work and didn't hold much grease got three shots every three hours or so. Slow-running bearings were greased daily on the morning service. It was important to understand what you were greasing; over-greasing the slide on a variable-speed pulley might end up with grease on the belt. On the other hand no grease would result in the pulley being unable to self-adjust and a ruined belt would be the outcome.

Dale was very strict on proper servicing and knew

which members of the crew he could trust, and whom he needed to keep an eye on. He talked of 'Grease Worms': I asked him to explain. He said if a bearing had failed and when removed was dry and devoid of grease, the operator would maintain that he had greased it regularly. Dale's reply would be that 'In which case it must have been an invasion of the Grease Worms; they must have eaten all the grease!'

Chains were oiled daily, Dale saying that he could tell at the end of the season by the amount of chain wear on a machine who had and who had not kept their chains oiled. We changed engine oil every hundred hours. The older machines did not have an hour meter so there was a degree of guesswork involved, but a week's work would often mean that a change of oil and filters was called for. Waste oil would be drained into a bucket then tipped out onto a dirt road. Even in those days, when 'environmentally friendly' was a term unheard of, this practice seemed dubious. However, it soon mixed in with the dirt and Dale pointed out that blacktop (tarmac) contained similar products and that there was a lot of blacktop in the USA.

Truck drivers would return from town with tales of the latest 'honey' working at the elevator. Many elevators hired high-school students, female of course, drop-dead gorgeous, and wearing the skimpiest of shorts and tops, to weigh the trucks in. A pleasure denied to us poor combine drivers, but I was still not tempted to swap jobs. Later harvests would see me working alone sometimes. I would combine a truckload of wheat, then haul that load into town myself, but this was rare – stopping a combine was the last thing to be done.

Tandem trucks held about six hundred bushels of wheat, approximately fifteen tonnes.

Loads were tarped (covered with a tarpaulin) for all road journeys, to prevent wheat being blown off the top of the load when travelling at speed. In the '70s rollover tarps were few and far between, so all our trucks were tarped manually. This involved climbing on top of the load and unrolling the tarp from the front of the load to the back, then climbing down to the ground and securing it with rubber tarp straps: not the best of jobs in those temperatures. It was usually dry enough to run the trucks alongside the combines to unload on the go. Only for the last few bushels did the outfit come to a halt to top the truck off. Any combine driver worthy of that title could top a load off without the need for the truck driver to scoop shovel it level.

We hoped to be cutting by 9.30 or so, by which time it was usually already unbearably hot, but of course the weather dictated this and it might be sooner or later. The heat seemed to be relentless. A chrome wrench laid on the combine in the mid-day sun would be too hot to pick up after ten minutes. In contrast to later years, those combine cabs, apart from the new one, were pretty basic. Thankfully, air conditioning was standard.

I was – and still am – always thrilled by the prospect of a day's combining. I never saw it as a chore. You are master of the machine which gives you information, via eyes, ears and nose, and you react to that information. Dale said you needed to be able to operate it through the seat of your pants, and once again he was right.

It became clear that all was not well on the domestic front. Jerry was doing a good enough job, but Dale would expect her to anticipate if we moved fields during the day, so when she brought out the evening meal we could be five miles from where she last saw us. Also Dale could not see the need to give her regular sums of money for the grocery bills. These were, in his eyes, mere trivial things that Margie would have sorted out.

We went into the trailer one morning to be met with a definite lack of breakfast. Words were exchanged between Jerry and Dale, and the crew loaded into pick-ups to head for Iowa Park café. We said a sad goodbye to the Costellos the next day and ate breakfast in town from then on. We bought bread, ham, etc. in town and fixed our own lunches. Dale took orders for evening meals, collecting them from the café in individual containers, and brought them to us at the field.

Apart from the truck engine, we had no major breakdowns in those first few weeks. A table drive-shaft sheared on my machine, but it was a pretty straightforward job to replace. The oldest combine, on its fourth harvest, gave enough trouble to convince me that three years was long enough to keep a machine.

As combine drivers, we were issued with a set of Challenger wrenches which stayed with our machine. The parts truck carried a comprehensive selection of wrenches, sockets, bearing pullers, nuts and bolts for use by everybody. We also had half a truckload of spare parts with us. There were truck tyres and tubes, combine belts, chains, axle stands and a large metal chest full of bearings and other precision-machined parts necessary to keep complicated machinery in running order. This lot had to be loaded into one of the grain trucks each time we moved states.

★   ★   ★

In those early days there was a lot to learn, both in procedure and terminology. I already knew that petrol was gasoline or more commonly 'gas', what I didn't know was that a combine knife was a sickle, a knife finger was a guard and that a guard was a shield!

We pulled onto a new field, Big Hieserman, a 150-acre chunk of wheat surrounded by other wheat fields, pasture and scrub. No farms or houses within sight – and you could see a long way. I happened to be first to pull onto the field and by then knew the routine. The entrance was part-way along one side of the field so I engaged the threshing mechanism while still on the dirt road and cut my way into the field. Cutting twenty yards straight out into the crop I then looped round and came back to the side of the field to follow the fence row. With the benefit of a straw chopper this exercise could be done quite neatly, no wheat was run down or wasted and the following combines had room to enter the field and follow me round.

'Who's in that front combine?' asked Dale over the radio from his pick-up.

'It's Robert,' came my reply.

'Well, cut east till you hit that first terrace then follow it south and come back west again.'

The only terrace I could think of was a row of grimy coal miners' terraced houses in Derbyshire, and a mental block prevented me from thinking straight. Five combines pushing up behind me didn't help. I couldn't for the life of me see a sign of *any* dwelling let alone a row of them.

This was my first encounter with terraces. Basically they are an earth bank formed by a land grader and follow the contours of the field. They are created to stop soil erosion by heavy rain or strong winds. These banks are cropped with wheat in exactly the same manner as the rest of the field. Some are quite low and can be cut over easily but most have to be cut along either side to ensure that all the crop is gathered and no dirt is dug up by the combine header. There were just two terraces in this 150 acres, and most of the Goetze fields were terrace-free. I was learning fast!

Dale conducted all operations from his pick-up: checking, advising and instructing both combines and trucks. He could tell how a combine was performing just by looking at it as it went past. 'Six-twenty to Robert' (six-twenty was our radio call sign).

'Go ahead' or 'Come on' would be my response.

'What is your cylinder [drum] speed?'

'It's 900, Dale.'

'Well, drop it to 850, you've got a heavy chaff condition.'

As the day wore on the wheat would become drier and the straw would break up. It would take the path of the grain, overloading the sieves, instead of taking the correct route for straw – over the straw walkers.

If all was running well Dale was also very good at taking a nap. A repeated 'Six-twenty to Dale' might eventually be answered by a sleepy 'C'marhn.'

Late one evening, Dale said I should take my combine back to Oscar's yard and clean it down. Apparently, I was to go north back to Oklahoma to help Dale's son Larry harvest the Starks' wheat. Cleaning that combine down was perhaps the worst job I did in five years of harvesting. The Goetzes had wild oats, and Dale did not want them to be taken onto his farm.

Although dark, it was still hot and sultry, and while the rest of the crew carried on cutting, I spent a miserable two hours cleaning those miserable oats out. To enable me to see what I was doing I had to use a lead light or 'trouble light' as it was known, which of course attracted every damn bug in that part of Texas.

The Fields of Oscar and Ernest Goetze

Oscar and Ernest lived about 5 miles apart

Map Not to Scale

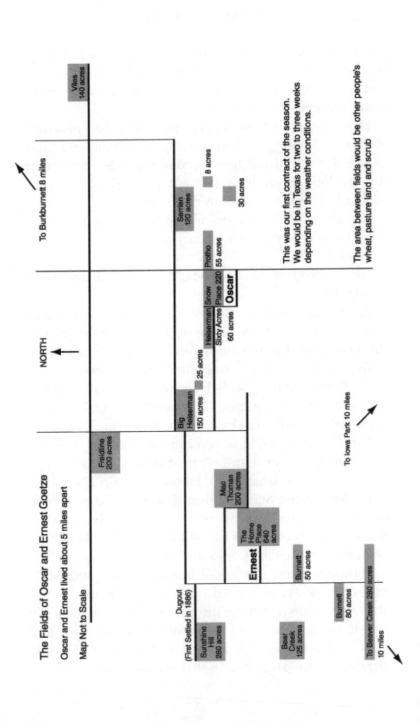

This was our first contract of the season.
We would be in Texas for two to three weeks
depending on the weather conditions.

The area between fields would be other people's
wheat, pasture land and scrub

Use of the airline resulted in my being covered all over with thick black dust which stuck to my sweat-drenched body. The itching was very unpleasant, I could have wept. I was still at it when the rest of the crew came in for the night.

I had discussed the wild oat problem with Oscar and Ernest. Sprays were sometimes used, usually applied by aircraft, but not to the extent of most European farms. Simply growing continuous wheat year after year made the weed problem worse. Ernest agreed that crop rotation would help. He said years ago they had raised cotton, but it was very labour-intensive and a heart-breaking chore in all that heat, so much so that all the cotton he wanted to raise now was in his girlfriend's nightgown!

The following day Larry turned up and we headed north for Manchester, Larry driving the truck pulling my combine, me driving his pick-up. Although I was sad to be leaving a crew of which I had become part, this was more excitement and a new adventure.

Larry was great to work for; easygoing once he knew I was putting my heart and soul into the job, but I suspect not one to suffer fools gladly, and not a man to cross. He had, of course, been brought up on a combine harvester: as a baby in a crib beside his mother while she drove a machine, helping her to operate one from the age of five, and in full control of his own machine by the age of ten. He had seen a lot of harvests, and a hell of a lot of harvest hands. I wondered what he would make of a green Englishman.

Larry referred to Dale as the 'old man'. There was a certain amount of tension between them, as you would expect from two single-minded men who happened also to be father and son, and who had been through so

53

much together in all those harvests past. Larry was now burnt out on harvest, as he put it, and concentrated on the farm at home.

Larry had collected one of the other 760s a few days earlier which he drove himself while I operated my own oat-free machine. We worked well together. Larry's wife Connie drove truck, hauling the wheat to the elevator in Manchester. I ate breakfast and supper in the house with the family; Connie would bring a sack lunch out to the field. I felt very welcomed by the family and that my efforts were valued.

By 9 pm or so Larry was ready to shut down. The wheat by then had got 'tough' and could be heard rattling against the straw chopper hood. We would go home, drink a beer and eat supper. Larry's attitude to beer was somewhat more relaxed than Dale's. Dale had hired many an alcoholic in his time and seen them wreck equipment as a consequence. Although he was known to drink himself on occasions he was vehemently opposed to the crew drinking, and I could see his point.

Some crew members were quite irresponsible with any aspect of alcohol and, given half a chance, would drink themselves stupid. I am not a big drinker, certainly not during the heat of the day, and not while operating a machine, but one cold beer at the end of the day did no harm . . . and we would visit bars as the season progressed.

As in Texas, in that part of Oklahoma, just one mile south of the Kansas border, the main crop was wheat. Apart from some land being cut for hay, Larry seeded the whole farm every year to wheat. There was less scrub than in Texas, and a good collection of trees and bushes to break up the landscape. Like Texas it was relatively flat. Dirt roads ran parallel or at right angles to each other at one-mile intervals to give one section of land, 640 acres.

As in a lot of the Great Plains, individual fields were often referred to by the original homesteader's name, so you might be directed to the Johnson Place, when in fact Mr Johnson and his family may have gone broke and left three generations ago. If you were lucky, as indeed we were further north, you might have come across the old derelict homestead and barn, rusting horse-drawn implements and perhaps a couple of cotton-wood trees that might have been the only trees around for miles. A veritable oasis in the desert.

When the Goetzes had talked of plowing their land, I realised they were talking about what we would term cultivating. After harvest the wheat stubble was worked with 40-foot cultivators, then in September seeded with an air seeder. This was a hopper on wheels holding about five tonnes of wheat, behind which another 40-foot cultivator was towed. A metering mechanism and a fan propelled the seed along pipes to the tines of the cultivator and into the ground at the required depth, spacing and seed rate.

In Oklahoma, Larry mouldboard-plowed his land. Two trailed five-furrow conventional plows were hitched one behind the other and the land was blacked over. Skimmers, accessories on the English ploughs we per-nickety Europeans use to completely bury stubble and trash, were noticeably absent. It was important to get the plows in as soon as possible, because once the crop cover was gone the sun would quickly bake the ground too hard to make plowing possible.

So after a few days harvesting, John, a local guy, came in to plow the land behind us with Larry's articulated John Deere 8430. There were no straw swaths to bale and haul; all combines had straw choppers fitted. Most years we chopped every acre of straw; occasionally, if we were working near a feedlot, we would pull back our straw choppers, and the straw would be utilised for the beef enterprise.

By late June, Larry and I had the Starks' wheat safely harvested. Dale and crew had finished in Texas. Our next stop was to be Leoti (pronounced Leota) in western Kansas. We all met up in Manchester where it was great to see everybody again, lots of handshaking, laughing, and stories to tell. We had lost another crew member. Dennis, who had been a self-styled foreman, and who I thought would feature prominently in the crew, had gone; though thinking back I remembered he had not been allocated a combine on that first day's cutting in Texas.

It became apparent that I was to haul my own combine. Somewhat daunting, but it was good to be given that trust. I had driven farm trucks as a sixteen-year-old in England on the farm where I used to work. But that was in the fields. This one was left-hand drive, on strange roads, through strange towns, and had ten tonnes of combine behind it on an un-braked lowboy.

Although I did not have a heavy goods licence, some quirk in the American system allowed me to drive a farm truck – as these were classified – on a standard international driver's licence, a loophole which has since been closed. I ought to remember which truck I drove, but am not sure. Certainly it would have been one of the older ones, probably the '68 drag axle, so-called because only one of its two rear axles was powered. It was a thrilling experience, one that never lost its thrill, even after five harvests.

All I had to do was keep the combine in front in sight. Easy enough most of the time, but tricky in towns where stoplights could intervene. From Manchester we went over the state line into Kansas, north through Anthony, west to Medicine Lodge then north again to Pratt. By now it was getting dark so we pulled into Kinchloe's Inc., the local Massey Ferguson dealer, and parked up for the night.

Then it was all aboard a couple of pick-ups, back into town to eat. Riding in the back of an open pick-up was always fun. Even after dark it was still hot, and this was a great way to cool down.

No shower that night, I was back in my old bunk, the one immediately above Charlie, in the 'select' end of the bunker. Breakfast in town the next morning, then we hit the road again. North on Highway 281 to Great Bend, Dale giving advice over the two-way radio: 'Six-twenty to all combines, go easy over those railroad tracks in Great Bend, they are real rough.'

KRV 620 was our call sign; this was business-band radio, not citizens band. You were supposed to use your call sign each time, using the airwaves sensibly with no unnecessary chit-chat. The authorities were known to listen in sometimes, and you could lose your licence for misuse.

Dale's Chevy pick-up had a mobile phone, quite something in 1979. It was hooked up to the loud horn when it rang, a little disconcerting if you were driving it through town, and even worse if you were underneath the engine changing the oil filter, and it went off – and yes, it did happen to me.

We turned in Great Bend, over several stop signs, and were to head west on Highway 96. I knew this was an important turn and was very anxious not to miss it. The stoplights were doing all they could to screw things up for me, but at last there was the sign for 96. I got into the right-hand lane and having taken the exit and rounded a bend, was greatly relieved to see the rear end of a 760 in the far distance. I relaxed a little now; getting the feel of the truck, and knowing that this highway would eventually take me to Leoti 140 miles away, even if I did get split up from the rest.

Our convoy of seven trucks pulling seven combines, the parts truck pulling the bunker and two pick-ups pulling two travel trailers must have been a nuisance to local traffic and people, but I never sensed any hostility from other road users. Most people had at least some vague connection with farming, and though we were rough and itinerant, many looked on us as valued members of society, doing the important job of gathering the harvest.

Jim Roland's yard, just north and east of Leoti, was to be our base while in Kansas. Jim was manager for Franks Farms where Dale cut wheat in the summer and corn in the fall, and had done so for many years. Trailers were parked and hooked up to water and electricity. The yard held several 'treasures' from harvests past: a '68 Chevy pick-up, an old trailer house, combine spares and attachments, all belonging to Dale. I felt a strong sense of history. What crews had operated this old equipment? What were their names? What had harvest meant to them?

We had four customers over an area of about thirty miles. I was part of the crew that cut for Carl Downes, fourteen miles south and three miles west of town. Kansas presented a new landscape: flat, very few trees, but fields of golden wheat were contrasted with fields of green corn. Some wheat land and all the corn were irrigated. Centre-pivot irrigation was not common in this area at that time, instead the land was flood irrigated. The land was left corrugated, that is grooved as if culti-vated with a wide-tined machine. A trench was created on the highest end of the field by a large tractor pulling a deep ridger.

The trench was kept full of water by a pump drawing water from many feet below ground. Then aluminium tubes shaped a bit like a boomerang were set out along the trench at regular intervals to siphon the water out of the trench and down the corrugations. To get the siphoning in action you had to thrust one end of the pipe into the water, put your hand over the other end keeping it below water line, thus trapping the water in the pipe, repeat the process a few times and, Hey Presto, the siphoning would begin. This was a job usually undertaken by casual labour. Jim said there was a definite knack to it, and intelligent people could not do it . . . I never tried!

Engines running on either diesel or natural gas pumped the water – water that was cold, clear and delicious to drink in the Kansas heat. The corn standing five or six feet tall often obscured the pump. If you needed your water jug replenishing, you simply had to get away from any combine noise and you could usually detect the sound of the pump engine roaring away. There was always a small tap in the outlet pipe to access what you were after.

Cutting these corrugated fields called for a different approach. Dry land (non-irrigated) was usually cut round and round, starting at the field boundary and working into the middle. Headers were not raised at the corners, nor did they leave the wheat. This left small triangles of uncut crop on the two diagonals. These

59

'corners' were harvested last. The idea was to keep the combines cutting at all times. It would not work with our heavy European crops but worked very well in those conditions, no wheat being lost.

Corrugated land was too rough to cross at right angles, for the land by now had baked dry. This meant working in lands in the same direction as the corrugations.

Irrigated wheat was yielding about 80–100 bushels an acre, dry land back in Texas made about 20–30 bushels. (One bushel is approximately 60 lb or 27 kg.)

So in round numbers a 40-bushel crop was yielding one tonne to the acre.

These western Kansas fields were uninterrupted by trees and patches of scrub, so tended to be a full section of 640 acres with four straight sides each a mile long, or a half- or quarter-section. If you pulled onto a full section of land at one corner and cut all the way round you would have travelled four miles. A truck in the same field as the combine might have to travel two miles to reach that combine if it was at the opposite corner.

We ate good food in Millers café in town, frequented by other custom harvesters in the area. There were two waitresses, one very attractive college student who Charlie got to know rather well and one dragon. We soon sussed that the restaurant was split between the two and it was always our goal to be sitting on the side taken care of by the pretty one! At the end of your meal you would sign the ticket with both your own and Dale's name, the bill would eventually be paid by Dale – and a pretty sizeable one it would be with a crew of ten or so eating two meals a day for perhaps three weeks.

Millers café was a typical wheat-town eating place. They put themselves out to accommodate custom harvesters during the harvest period, and if crews ate there regularly would not close at night until all had been fed. Some cafés displayed notices 'Custom Harvesters Welcome', some were reputed to increase their prices during harvest when they had a captive audience, but it could be argued that this was their harvest time as well. If we were in a strange town and there was a choice of eating places, Dale would note where the truckers were eating and we would follow their example.

Gary found a girlfriend and quit shortly before we left Kansas. I was sorry, I liked him. How experienced he had seemed to me as he had driven me south to Texas on that first haul, and how little I knew, and now within two months Gary was quitting, demoralised by the whole thing.

Two new hands, Rick and his mate Glen, joined us. Rick was very amiable, and made a good truck driver. He adopted the '74 tandem, lavishing lots of care and attention on it. Glen – we called him Griz because of his similarity to a grizzly bear – was equally amiable and took on one of the combines which he called Sabbath.

★   ★   ★

Our next stop was eighty miles to the north-west at Cheyenne Wells, Colorado for a Mr Smith. Again the landscape changed a little, becoming less fertile and less prosperous than Kansas.

We had some weedy wheat to cut. Fireweed, green and succulent, especially after heavy rains, would jam table augers. You would have to leave your comfortable air-conditioned cab and unplug it manually in the hot sunshine.

It was while kneeling down in front of the header on one of these sorties that I heard, then saw, a rattlesnake not six feet away from me. I kept a close eye on him, unplugged the auger and beat a hasty retreat back up the ladder to the cab. In open ground with a suitable stick I would have tackled him, but in standing wheat with lumps of the debris I had just relieved from the auger, and the combine in the way, I thought he might out-manoeuvre me. At the shout of 'snake!' I have seen grown men leap into the nearest vehicle, even locking the door – surely the reptiles are not that clever? Rattle-snakes, potential killers that they are, did not bother me unduly, but I was much more of a wimp where rats were concerned.

Apparently it depends on your metabolism as to how a snake bite affects you. The secret is to stay calm and then the poison takes longer to get into your system. Small comfort if you are out in the sticks and fifty miles from the nearest antidote injection. I thought back to what Ernest Goetze had said to me in Texas: 'If you have a broken rubber in one pocket and a rattlesnake in the other, don't fuck with either one of them.'

Because Colorado had more scrub and bush than Kansas there was a little more wildlife to see, but not much. By now I had experienced the wildlife of four states. Mostly it was snakes and small mammals. The occasional deer would run out of the wheat, as would skunks. Armadillos, racoons, possums and turtles were about, but unfortunately often seen as road kill. Coyotes were heard howling at night rather than seen by day, but occasionally the odd one would slink by. We saw jack rabbits – they looked to be more like a hare than a rabbit – while rabbits were known as 'cottontails'.

There had been plenty of mosquitoes so far, especially near water. Evenings were the worst, there was every incentive to stay in the combine cab and cut wheat; the truckers suffered most, waiting around for their truck to be loaded. A few birds flew around and sang to us when lack of machine noise made this possible, and I did identify a pair of quail as they took off from the Colorado scrub. What I loved to hear was the sound of the crickets late at night when our combines were silent at last.

One afternoon a straw chopper bearing got hot, and a piece of red-hot metal dropped on the ground setting fire to the stubble. A cry of 'fire!' went out over the radio, and a frantic beating of flames by all those near took place. It was out in minutes, with no damage to the crop or vehicles. I was over the other side of the field and not involved. Wheat fires were dreaded by farmers and custom harvesters alike. Some farmers would have a tractor and cultivator or disc hitched up ready to plow firebreaks across the wheat or stubble if a fire broke out.

Even a small town like Cheyenne Wells would have a laundromat. A wet day would give us the chance to get some clothes washed. In long periods of dry weather you would end up sorting through dirty clothes for something slightly cleaner than you were wearing. This got to some people; and although it did not bother me too much, I have never taken a shower and clean clothes for granted since.

When the weather gave us no break for several weeks, Dale would detail a lowly member of the crew to take everybody's laundry to town and get it clean. A thankless task that fortunately never fell to me.

Doing your own laundry was, however, quite a social occasion. A chance to get into town, write some letters, hang out with the crew and check out the local girls. I would also take the opportunity to catch up on some sleep, and might be found in the seat of the pick-up snoring away.

Slim's drinking was becoming a problem. One night he took off in the Caddy in search of beer, got drunk, got the Caddy stuck in a field on his return, and was sacked. Our one experienced man was gone, but by now the rest of us were beginning to feel like veterans.

Eventually our weedy wheat was cut, and as our next contract, in Montana, was not yet ready, Dale found us some more cutting, 150 miles to the north-west, to fill in the gap. With the weather having such a great influence on our timetable it was nigh on impossible to dovetail jobs exactly.

Dale had a nucleus of regular customers, and would take on or refuse extra work according to the season. The next job could be ready and the current job not completed, which might mean splitting the crew to keep both jobs. It must be very hard to watch it pour down with rain on an unfinished job. You are paying a non-productive crew, knowing that your next job is ready to cut, the weather up there is fine, and that your customer is surrounded by other cutters looking for work.

Gary Lindenvok of Byers, Colorado grew 26,000 acres of wheat without possessing a single combine harvester of his own. We were one of ten cutters working for him and, not being one of his regular cutters, got landed with some pretty rough work. More weeds; bad enough that we simply cut round large areas where the weeds prevailed and the crop was virtually non-existent. The land, some god-forsaken corner of Colorado, was fifty miles from the nearest settlement. The story went that even the rattlesnakes took a sacked lunch!

Still, it kept us busy, though how Dale worked out with his customer exactly how many acres to charge for I know not. You would never get specific information on such matters out of Dale. Any inquiry as to how many acres we had cut or had left to do was met with very vague answers. Dale would have made an excellent politician.

We made camp in the small town of Byers and ate in the café. On Sunday morning we found it was closed and it looked like we might be giving food a miss that day. Somebody obtained coffee from somewhere; being a non-coffee drinker my breakfast consisted of a can of Cola out of a machine. My stomach complained bitterly at this abuse.

Isolated places such as the one we were cutting out at this time were often referred to by the crew as 'Bum-fuck Egypt' or 'Screw-you Peru'. I'm not sure who coined these phrases but they always raised a smile.

Terry Pfeiffer and his crew, Mark and TR, were with us at this time. Down south he had worked for his own customers for part of the time. In his late twenties, Terry had been cutting for a year or two on his own. Dale had seen his advertisement for prospective cutting and invited Terry to run along side him with the possibility of eventually taking over the business. As if Dale was going to retire! I'm not sure who was kidding whom.

Terry was quietly spoken and easy to get on with. Next year he ran on his own again, even taking our Cheyenne Wells customer – but as Dale put it, 'I cut those weeds till my ass hung out . . . and it bled!'

Terry Pfeiffer and I were to take two combines to Montana, a journey of about 850 miles. Dale announced these things; there was no discussion or consultation. I was happy to fall in with whatever was asked of me.

With Terry driving his Ford truck pulling his combine, I pulling my combine in, I think, the drag axle, and

Terry's girlfriend Suzy driving his Datsun pick-up, we headed for Montana. Little did I know then that this would become our favourite state, that we would make great friendships here that would last a lifetime and that the times spent in the Triple T Tavern would become legendary. This was all to come, but for now we were heading for a town called Gildford, in northern Montana, on the Hi-Line west of Havre. We were to meet up with a guy called Levi Hanson who would lead us north of town to his land we were hired to cut.

West to Denver, then north on 25 through Cheyenne and Casper to Sheridan, Wyoming. Sheridan was a port of entry. We were required to check in with the authorities there to pay our state taxes for the privilege of using their roads. All states had them and you paid according to your circumstances, perhaps a thirty-day permit to haul grain, or in this case a permit to haul two combines across the state.

As Sheridan was on the northern border of the state and we entered from the south, Terry would have to ring into the port of entry office to state our intentions. Police patrols might take your licence number, call through to the port office and if they had no record of you, there would be trouble!

We never cut in Wyoming and I saw very little wheat there. This was cattle country and pretty rugged too, a marked contrast to the flat lands of Kansas. I noticed that although the days were hot, it cooled down at night and you could breathe. I remember the absolute luxury of needing to put a coat on in the evening for the first time in ten weeks.

From Sheridan, with Terry leading, we travelled north to Billings and on to Roundup then headed for Harlowton and Moore. We pulled into Fort Benton at dusk and shut down for the night. Terry checked us in at a motel, one room for me and one for Suzy and himself. After a meal in town we hit the sack. At breakfast the next morning I got the feeling that I'd had more sleep than them!

After breakfast we walked round town. I bought some postcards and presents for family at home and for at least half an hour behaved like what Dale would call a 'goddamn tourist'.

Even an innocent visit to town such as this might be used by some crew members to buy beer and get drunk. I could see why Dale discouraged such behaviour and was reluctant to hand over any great quantity of money at any one time, knowing some people might end up being incapable of doing their job or, worse still, tearing up machinery.

Slim had said he was 'broke and out of town'. He also said he was 'so hard up he couldn't pay attention'. The crew had by now formed the Seldomites Club, so-called because we were seldom in town, seldom had money, and seldom saw women.

From Fort Benton it was a short haul to Gildford where we arrived early afternoon. I liked the look of this small town with its typical layout: railroad tracks bisecting the main drag, small elevator alongside, gas station at the co-op, wooden houses, a post office, a café, bar and store, with grain bins scattered amongst it all. I could never have guessed that this place would become very special to us.

We pulled up just short of the railroad tracks and parked on the side of the road. By sheer chance we were almost opposite Levi's house. As we were checking tyres, drinking Kool-Aid, and generally feeling pleased with ourselves for making it in one piece without too much trauma, a red pick-up pulled up and inquired if we were Starks' crew.

Levi Hanson was in his thirties, had a moustache like the Sundance Kid and struck me as rather different from the average American farmer. He led us north on dirt roads about eight miles where we unloaded and started to cut his wheat.

The rest of the crew could only have been a day or two behind us as we only stayed in a local motel for one or two nights. They eventually arrived and made camp south of Kremlin at the farm of Norman Dack. This would be about twenty-five miles from where we were cutting so as soon as they arrived I went back to my old bunk at night.

After a couple of days Terry took his machine to another job, leaving me to cut alone for Levi with a young kid hauling. Levi introduced me to his wife Dian, who brought out food to the field for us. They were a lovely couple who were both educated and had led a carefree hippie life in North Carolina after leaving university. Dian had taught school part-time for a small income, they lived in a cabin in the woods, skinny dipped in the lake and, I suspect, smoked a little weed.

They had returned to Montana to take over their uncle's farm. They farmed about two thousand acres organically and, in that dry part of Montana where you might only get eleven inches of rain a year, were struggling to wrestle a living out of the land.

Though pesticides and fertilisers were used on the Plains it was to a much lesser degree than on most European farms. However, this was the first organic farm we had cut on. For Levi and Dian it was a lifestyle choice, they did not want to pollute the land – and their produce – with chemicals; they were also vegetarians.

My truck driver and I were asked if we would care to eat with them one evening. This was an almost forgotten luxury. To get cleaned up, changed into clean work clothes and eat at someone's table was quite something. We still cut till the dew stopped us, so there was no shutting down early, but to sit up to a table in someone's house and use knives and forks was a very pleasant experience. It made me realise just how rough we lived for most of the time. Not that I would have changed anything. I loved the life, all of it.

We had a lot of work in the area. Wade Reese, Dale's old friend, had a few acres to cut. Wade and his wife Edna made us very welcome. Terry Laws, Charlie and I would call in for tea and cookies on our way to the fields. Wade always joked that he baked the cookies, with a glint in the old-timer's eyes we knew it was Edna who was responsible. Wade would fix up work for Dale in the area, and tell us stories of harvests past. He had known Dale since 1950.

I loved the climate and the huge horizons of Montana. It was easy to see why it was called Big Sky Country. The days were still hot but the first chills of the approaching fall were evident some evenings.

My 1977 MF 760, now named Maria after a waitress in Wheatland, Wyoming, on its third year had given very little trouble over the season. The table drive-shaft had broken in Texas, requiring oxy-acetylene cutters to remove the debris, and a few other odds and ends had required attention. A new set of tyres was fitted in Colorado, the old ones being worn out, to give some idea of the amount of work it had done.

Hour meters were optional equipment on those machines at that time, and Dale chose not to have them. It meant guessing when the next oil change was due, but

Dale always erred on the safe side where oil and filters were concerned. Though he never said it, without an hour meter it would be hard for the machinery dealer to know just how much work the machine had done when it came to trading.

One day, when I was working south of Kremlin, the machine jumped out of gear and would not stay in gear without my holding it in place. I explained the problem to Dale who said we were too busy to stop, so to hold it in gear till the weather broke. The only satisfactory solution I could find was to tie a rope from the gear lever back to the seat support. I ran like that for a whole day.

When the weather broke that night Dale suggested I 'jerk that transmission' the next day. Most repairs were undertaken in the field, but as this was a sizeable job and there was a suitable shed or 'Quonset' nearby and available, I had the luxury of a concrete floor to work on.

A bearing had broken up, and running it a day had screwed up all the other bearings. I would not have tackled a job like this myself back in England; but with a shop manual, Dale on hand to give advice, and the philosophy of fixing your own problems, I took the transmission apart and replaced all the bearings therein.

As I was doing these repairs, Charlie, my British companion, was giving his combine a spring clean. I think she was cleaner then than when she had rolled off the production line, a credit to his efforts. The new combine had had one or two drivers those first few weeks in Texas. Terry had driven her for a while, but eventually Charlie took her over, called her Bessie and lavished a cleaning regime on her that would have done justice to an operating theatre.

Terry had been lumbered with the older 750, its driver having been unable to keep it running back in

Texas. Terry helped him out and ended up running the thing himself, thereby paying the price of being a good operator and mechanic. The main core of drivers would stick to their own machines, but occasionally Dale would take a new guy off a combine and put him in a truck and vice-versa. It was traditional that established combine drivers hauled their own machine when moving between jobs.

It was in Montana that we were able to stage a photograph of six combines working in echelon. Though it looked impressive to work like this, Dale discouraged it. If a machine plugged up or broke down, it held up the others; also, chopped straw and chaff ejected from the machine in front would block the radiator and air cleaners of following machines. He liked them spaced out around the field away from each other.

We had left the oldest 750 back in Colorado, Dale having thought it too old and unreliable to take to Montana. Even so, we ran the damn thing the next year, its fifth season, by the end of which just about everything short of the seat and steering wheel had been replaced – and they were getting shaky! It was my good fortune never to have driven it.

With Terry Pfeiffer's two machines we were able to field six combines and a grain truck for the photograph. Dale just about tolerated this. Terry Laws asked him over the radio what he thought of this impressive sight, to which Dale replied scathingly, 'Jesus Christ, a million dollars worth of equipment and we are taking pictures.' But I suspect he was as proud as we were of the scene.

Dale said he could not understand why somebody would want to take a picture of a bit of red metal. He once asked Charlie why he would want to take a picture of some mountains in Wyoming, saying that he had been coming past them for the last thirty years and they hadn't moved an inch!

My time was running out. It was always known that I would have to be back in England in early September to resume my fertiliser spreading operation, once the fields back home were cleared. We were just about done in that locality. Some of the crew were to cross into Canada, to the vast open spaces of Saskatchewan; the rest would go a hundred miles west to Shelby and Cut Bank. Most of this work would be picking up swaths, as this far north the wheat ripened unevenly. In Canada the crew would also be picking up canary seed, another important crop in that area.

Wheat harvest would be finished by the end of October then corn and milo harvest would follow in Kansas and Nebraska; that too would be complete around the middle of December. I would miss this later part of harvest and be sorry for it. Terry Laws and Charlie would relate their adventures of corn harvest to me, a time when temperatures would plummet well below freezing and snow would often disrupt work.

★   ★   ★

My plan had been to do just the one harvest, but by that time I knew that I would come back next year for sure. That I was able to share the following two harvests with Charlie, and the following three harvests with Terry was a very special privilege. Friendships and comradeships formed then have stayed with me ever since.

It was with a heavy heart that I packed my suitcase, said goodbye to the crew and farewell to my combine and the trucks I had got to know so well.

Dale drove me down to Great Falls, booked me in at the Ponderosa Inn where we shared a final meal.

It was getting late and Dale needed to get back to Kremlin to rouse a crew for their work the next day. He said thank you for all I had done, we shook hands and said goodbye. I flew from there the next day, home to good old England.

Chapter Four

New Combines

IN the winter/spring of 1980 Dale had traded the three old 760s for three new ones. Charlie, Terry and I had one each, Rockin Dopsie, Bogart and Marianne respectively. Dale hauled these machines himself from the Canadian factory in Brantford, Ontario with his White Freightliner and a hired low loader, a round trip of 2,500 miles for each combine. Charlie and Terry undertook the dealer's pre-delivery inspection themselves.

The cabs on these combines were much better than on my previous machine. They were quieter, more sophisticated and even had radios in them, music bringing a new dimension to our working day. It also helped with simple information, like what day it was! If we had a cook, and were not eating in town, with no newspapers, TV or radio there were times that first year when it had been impossible to know whether it was Wednesday or Sunday. The sight, sound and smell of those three new machines were a sheer joy.

I was late going out that year due to my farm management exam. June 6th saw me flying to Dallas then on to Wichita Falls to be picked up by Larry John, Oscar Goetze's eldest son. Fortunately harvest was late and had only just begun that morning. My brand-new combine had stood idle and unused all day in Oscar's yard, on Dale's instructions. That meant a lot to me.

We also ran Bessie and the old 750, now known as Loom Bow (rhymes with cow) because of the noise it was going to make when it finally blew up.

By now, Terry, Charlie and I felt like veterans. We had many hours' cutting behind us, knew the routine and felt passionate about our work. Terry had worked for Dale over the winter, overhauling machinery and helping out with cattle on the ranch in Nebraska.

After a new combine had completed its first season it would have a major overhaul the following winter in the Butler building at the farm at Manchester. The concave would be removed and taken to a machine shop to be hard-faced. Wear plates would be fabricated and installed in the feeder house (front elevator), clean grain auger and elevator before these areas wore completely through. Rasp bars would be replaced after about six hundred hours' use. Terry was good at this shop work and enjoyed doing it.

Dale's old pick-up appeared on the scene, the one featured in 'Yellow Trail'. It had been collected from Kremlin, Montana as the crew left that area in September 1979. It went all the way with us that harvest. Terry called it Bernice, and a couple of years later bought it from Dale. It would eventually exceed 450,000 miles travelled.

By now all the trucks had names. They each had their own character and it seemed right that this was recognised: '77 Chevrolet tandem (Mule Skinner), '74 Chevrolet tandem (Mo), '73 GMC Diesel (Jimmy), '70 Chevrolet twin screw (Humphrey), '68 Chevrolet drag axle (Dermot) and '67 Chevrolet four-wheel bobtail (Boris).

Dale did not enter the spirit of naming vehicles, but we took no notice! All but one of these trucks had appeared in 'Yellow Trail'.

The Goetzes treated us like old friends, and it was good to be cutting familiar fields, with names like Sunshine Hill, Bear Creek and Big Hieserman. The capped oil pipes, well-hidden by wheat in two of the fields, would be cut round by Terry, Charlie or me as soon as they were spotted, to avoid any nasty accidents by new crew members.

A year or two later when I was approaching one to cut round I was trying to decide which course to take. I could go just to the left of it or to the right or straight for it, stopping at the last moment, reversing back then cutting to one side.

I decided to go to the right but felt I was a bit too close so I dipped the clutch (these were manual transmissions not hydros). By a chance in a million the clutch linkage had become disconnected although it had been fine only minutes before. The combine carried on, I was not too close so I missed the pipe and eventually stopped the machine with no damage done. If I had decided to head straight for the pipe I dread to think of the damage that would have been inflicted.

We had a new cook. Donna, in her early thirties, had her nine-year-old daughter along with her and lived in the Terry Trailer. She cooked good food – so fingers crossed she would stay.

Talking of fingers, Todd, a new recruit, had his thumb almost severed as he was putting a knife back in a combine header. It was hanging only by the skin and he was taken to hospital. Todd didn't do much more harvesting after that.

Larry John Goetze drove a grain truck for us in Texas and it was with him and his twin brothers, Tim and Jim, that some of us travelled the 150 miles to Fort Worth one evening. We visited various dens of iniquity, ate a steak supper and got back to the bunker at five in the morning. Breakfast was at seven o'clock. We beat Dale by five minutes!

Road trips such as this, rare as they were, were only possible by using other people's transport. Dale would not have sanctioned the use of one of his pick-ups, we would not have even asked. Harvest has been likened to being in the army, where certain freedoms are denied. A likely scenario would be for the crew to have got blind drunk, wrecked the pick-up and be unfit for the next day's work. I would like to think that we were a little more responsible than that – after all we were there to cut wheat, it was not a holiday or a sightseeing tour. But our trip cost us no cutting time: we had seen some female flesh, eaten good beef, drunk good beer and felt the better for it.

Texas finished, we all moved to northern Oklahoma. Based at Dale's farm, we had several customers in the locality and were split up. Terry and I cut for Larry, Charlie for a local banker who also farmed some land. Terry and I were working fewer hours than the others. Larry shut down by 9 pm, saying the wheat was too tough. Late one evening Terry and I heard frantic exchanges over the two-way radio. The others were experiencing some kind of drama, but we had to wait for them to return to the bunker to get the full story.

Mule Skinner, our newest truck, had got bogged down in soft ground. A spark from the muffler had set fire to the straw and the whole lot caught fire. Dale was on hand to direct operations. When she was fully alight he encouraged Charlie to get in the cab and drive it out – I think the original driver was too scared – but to no avail. She was stuck fast. The cab and engine were completely burnt out, melting the front of the

77

aluminium box out also. An area of stubble was lost including some, but not much, wheat. It was a sickening experience but fortunately nobody was hurt.

The previous year another truck had been wrecked along with the John Deere tractor it was carrying when it ran off the road in northern Kansas, driven by Mike, an alcoholic, who was working on the ranch in Nebraska.

The wheat in western Kansas was ready before Oklahoma was finished. Dale said I was to load my machine and take it the 250 miles to Leoti. This was the first road trip hauling combine I had done on my own, so I was anxious to get it right. I was keen to appear to Dale that I had the confidence and capability to do it, but in reality was quite nervous. Dale gave directions: that I must go north till I hit fourteen. Fourteen what? I asked myself, elephants? Then I realised he meant Highway 14. At least I had a decent truck, Mo the '74 Chevy tandem. Dale had earlier been to Anthony, the nearest port of entry, to get me the appropriate exemption documents for hauling a combine and wheat in Kansas.

So, with my Kansas exempt sticker displayed on the windshield, a few clothes, sleeping bag and water jug in the cab, I set forth.

Leaving Manchester some time in the afternoon, I made it to Rush Centre by dusk. As instructed, I called Dale to inform him of my progress, and was relieved to hear that he was pleased with my efforts. Rush Centre did not appear to have a café, but a small store was still open, so I bought pies and cakes to provide an appallingly unhealthy and unbalanced supper. All part of the fun!

Sleeping in the cab by the roadside was more

excitement, I felt like a pioneer heading west. Most of the trucks had a bench seat and a person of my height and weight could get a fairly comfortable night's sleep. On the road again at first light, the last hundred miles were easy. Stopping in Leoti, I had a proper breakfast and bought a few provisions.

No mention had been made by Dale of where I would sleep or eat. The priority was to get that combine to Kansas and get it cutting. Domestic matters just didn't figure in his thinking. We were used to this, and it was all part of the adventure.

Carl Downes was pleased to see me. Luckily I had remembered how to find his place seventeen miles out of town – not easy with miles and miles of wheat and dirt roads, with few landmarks. Carl was a jovial character with a wicked sense of humour. In his forties he and his son, Carter, farmed about 1,500 acres. Some they owned, some they rented, and I think they were struggling financially. They lived in town, but there were two houses in the farmyard. Carl's dad lived in one; the other belonged to Carter, who only stayed there occasionally.

The wheat was ready so I unloaded the combine and, with the help of Carl and Carter, got the header out of the truck box. And so the wheels were turning again, I was lucky that my first big trip alone had gone so well.

Harvesting clean Kansas wheat was a joy, with no hills, no rocks and no weeds. Although it is good to work as a team, it is also good to cut a field on your own. You can do it to your own standard, cutting low and clean to achieve a tidy job.

On a few farms we were told not to cut low. High stubble trapped more snow and more snow was precious moisture. I hated cutting high, it looked so untidy. Carl drove Mo, hauling the wheat into town. He also provided his own farm truck to ensure the combine was not held up.

At 7 pm Mrs Downes brought out a meal for me. The car pulled into the field, and a huddle of people gathered at the end of the run I was on. Three hundred yards away I was cutting towards them. I had felt a little digestive trouble earlier, and now I realised I needed the toilet, and soon. What a dilemma! Not a hedge or cover of any description in sight, I was getting closer and closer to them. By then I knew I could not wait any longer, certainly not the time it takes to eat a meal and exchange pleasantries with a person you have not met before who has kindly brought you food.

I had to act. Stopping the forward motion of the combine, but leaving the engine and mechanism running flat out, I ran down the steps, went behind the machine, hopefully out of sight, and was only just in time. What a relief! The deed done, I casually got back in the cab, trying to make it look as though I had been adjusting something. I think I got away with it.

Carter's house was put at my disposal that night and for several nights to follow. Shower and bath, kitchen and bed, all to myself. Had Dale arranged this with Carl? . . . I doubt it.

In the mornings I would take a loaded truck from the previous evening into Collingwood Grain, the elevator in Leoti. Then it was into Millers for breakfast, and buy ham, bread and cookies from the store. As I had no service truck with me Carl would supply the diesel for the combine; this would be taken into account when the bill was figured out.

Charges at that time were about 12-12-12. That is twelve dollars an acre to cut wheat, twelve cents a bushel for anything over a yield of twenty bushels per acre (called overage or overrun) and twelve cents a bushel for haulage. For some reason the pricing system was different in Montana: in that state a fixed rate of fifteen dollars per acre was charged to cover both cutting and haulage. Of course, if large haulage distances were involved this would be charged extra.

In due course the others arrived. We had a lot of work in the area. Donna had quit, so we needed a cook. Benny Williams had also gone. He was a character, telling many a funny story in his heavy American accent. He had messed up on combines and had been put on trucks, where he also messed up. He suffered from asthma and the dust got to him. His breathing was so bad in the bunker one night I was convinced he was going to die.

We were all working long hours. There was always the urgency to get the crop harvested; the west is an area of sudden violent storms. Heavy rain or worse, hail, could devastate the wheat crop so that a year's work and income would be lost in minutes.

Dale found another cook from a nearby town. Kay was in her late forties and she too had a daughter, age about ten, along with her. They lived and cooked in the Terry Trailer that Donna had vacated. Dale had bought a new Chevrolet Scottsdale pick-up that year, which meant that the resident cook had use of a pick-up at all times. Our pick-up count had now reached a total of four.

We usually had a couple of header trailers along with us and an extra pick-up was always useful. It was quite legal to travel on the dirt roads and some of the minor blacktop roads with a 24-foot header still attached to the combine, and moving short distances between fields we often did. Sometimes a narrow access or a gateway meant that a header trailer was required.

A header trailer also saved the task of loading the header into the truck box on long hauls – provided that a pick-up and driver were available to haul it. If they were not, the header trailer with header on board was loaded into the truck box as one unit.

Some of the crew cut in Bennett, Colorado, but the rest of us were busy in Leoti until Montana was ready. That was a big haul, over a thousand miles.

John Fox, our new truck driver, had taken over the Jimmy Diesel. In his late fifties he was much older than the rest of the crew and did not fit in too well. Somehow he ended up hauling my combine, Marianne, on this road trip while I hauled some other combine. I was not happy about this. One got very protective of one's own machine.

Kay had been good help in Leoti but was uneasy about going to Montana; she had never been that far from home before. We did what we could to reassure her, cooks were important to us! No major breakdowns on that trip. A few flat tyres I expect. We probably averaged about 30 mph. Truck stops or port of entry would be a likely place to spend the night, Sheridan, Wyoming was one of them. Heaven knows what the fuel bills were. We only had one diesel truck; the rest were gas.

When we were on the road like this for several days the chance of a shower at the end of the day was unlikely as the bunker was not hooked up to water overnight. Three days on the road in that heat and you could start to smell your own body odour.

Gildford looked good, it was almost like coming home. Dale was worried that Kay was about to quit, and gave us strict instructions that we were to hook her trailer up to water and electricity *before* we unloaded the combines. Well that was something new: Dale putting domestic matters before cutting was unheard of.

We caught up with Wade and Levi, word soon spreading that Starks' crew was in town. A new customer, Charlie Griffiths, had hired us. Charlie Norman and I took our machines fourteen miles north of town and spent quite a time up there.

Unable to settle, Kay quit after a few days and caught a bus back to Kansas. Fortunately Gildford had a café, and it was there we would eat breakfast. By the time we got in at night it was closed, so it was over to the Triple T Tavern for a pizza. The tavern was run by the Toner brothers: Tom, Dave and Jerry. They farmed some land and also had a repair shop.

Charlie and I acquired a cooler. We would buy provisions in the store after breakfast, and during the day Charlie would call me up on the radio – we were in the same field – and invite me over for dinner. I would ask if dress was formal to which he would reply, 'No, just come as you are.'

Indeed, I was obsessed with food. We worked long hours and occasionally, apart from breakfast, saw no food all day. It did us no lasting harm but it sure did make you appreciate a meal. It also sharpened your improvisation skills.

Dale would be on his way to you with a hamburger which might arrive stone-cold by the time he had been round the rest of the crew twenty miles away and maybe fixed a breakdown or two. I started carrying tinfoil so that I could wrap the cold burger and place it on the exhaust manifold alongside the turbocharger. Fifteen minutes of cutting later, supper would be ready, piping hot.

It was at this time I was introduced to the Hutterites, a religious group living and working in closed communes of about forty-five people throughout those northern states and into Canada. Their origin dates back over four hundred years. German is their first language and they speak English with a German accent, despite having lived in America all their lives. The community close by where Charlie and I were working farmed ten thousand acres, five thousand of which would be into wheat at any given time.

Unlike the Amish they operated tractors, trucks and pick-ups extensively on their farms. They had good up-to-date equipment, but still could be found at the rubbish dump scavenging anything useful. Living in a closed community they only married among themselves, and though they would ship people around from other colonies, on a typical commune there would be only three surnames.

Everybody worked hard; nobody was paid a wage. You wanted for nothing. Families would live in their own house on site but all ate communally. The women cooked, the men eating first. All farm profits were for the benefit of the community. When a commune reached 40–50 people it would split and finance a new commune.

They were buying land all the time, putting parcels of land together, sometimes paying in cash. They paid state taxes but educated their children in their own schools. Discipline was very strict – no alcohol, no television. A car or pick-up was available for trips into town, usually to sell their garden produce, home-made bread, cream and other provisions, but even basic access to the outside world was severely restricted.

Local opinion was mixed. The Hutterites were

friendly, but some people did not like to see them buying more land – land that would never come back on the market again.

Some of the men would visit us at Charlie Griffiths' place. They were friendly and easy to talk to. You might be having a conversation with one guy, thinking you talked to him yesterday, then realise it must have been somebody else. Most of them looked alike. They all dressed the same with a checked shirt and black trousers held up by suspenders (braces). On their heads they wore a black broad-brimmed hat, or baseball-style hat with its advertising logo removed. The women wore long dresses and head scarves.

That they looked after their own, sick and old alike, must be applauded. However, they were secretive. Apparently, boys around the age of sixteen disappeared for a few months. Any enquiry as to their whereabouts was met with lame excuses: 'He's in town' or 'He's busy in the fields'. Even those who left the community, and these were very few, would not speak of their way of life.

Dale had sold his last two Massey 510 combines to a commune, not this one, down by Great Falls at the end of the 1975 season. Margie, Tom Kirk and another hand had delivered them.

I needed a pulley welding and, as it was fourteen miles back to town, Charlie Griffiths suggested I take it along to the 'Hoots' as they were known locally. With some trepidation I drove into the commune. It was quite busy with people of all ages and both sexes going about their daily chores. I asked for George or Josh Vortz, and, explaining the reason for my visit, was directed to the workshop.

And what a workshop it was. They had all the latest equipment for repairing farm machinery. The man in charge soon had my pulley fixed, and then said in a heavy German accent, 'You come eat with us now.' I wasn't too sure about this. Was I about to be abducted? White slave trade?

The dining hall looked pretty much like a school canteen with long trestle tables and wooden benches to sit on. The women busied themselves and platefuls of food arrived. I was sure about the chicken but not too familiar with the rest of it – though it was all good to eat. I sat with the men almost opposite some teenage boys. The boys found my accent very amusing. Politely trying to stifle their laughter, they obviously had not come across an Englishman before.

The colony seemed well established, the neat, tidy houses and gardens looking like they had been there several years. I was surprised to learn that they did not have mains water on site but hauled all their drinking water by truck from town, twenty miles to the south.

Eventually it was time for the women to eat and for me to go. They had been helpful, polite and hospitable.

Most of the land we cut in that location was reasonably flat but there were a few steep areas. We had not cut for Charlie Griffiths the previous year so those fields were unfamiliar to me. I pulled onto a new piece and started cutting round the edge of the field clockwise in the usual way. The field had a watercourse, this one dry, known in these parts as a coulee (rhymes with Julie), on my left, a twenty-foot drop off the edge of the field. It was always nice to have a feature like this when you had cut strips for days on end. I followed the field boundary along the coulee enjoying the view.

By now my grain bin was getting full and that meant the centre of gravity of the combine was higher. Those combines were heavier on the left-hand side; most of the drive components were that side, indeed the left-hand drive wheel had ten pounds per square inch more air pressure in it than the right, to compensate for this. Looking ahead I could see the land starting to slope down to the left, and this slope got progressively worse.

I started to get a bit worried and realised I was doing this all wrong. I should have been cutting this slope in the opposite direction with the heavy side of the combine uphill and with an empty grain bin. The combine was at an alarming angle, the cab offset to the left and I was looking down a twenty-foot drop into the bottom of the coulee. I bottled out, swung the back end of the combine round and headed safely up the slope out into the crop.

I radioed Charlie to put him in the picture and while I went for a change of underwear he came at it in the opposite direction with an empty bin and cut the slope out. With a mental picture of a 760 combine, less than a year old, upside down in the bottom of the coulee I thought back to something Dale had said in *Yellow Trail from Texas*: 'Where the grain drill hasn't been you have no business with a combine.'

I named that field Kamikaze Coulee which gave Charlie Griffiths a laugh. Revisiting that field twenty years later I was rather disappointed to find that the slope was not as steep as I had remembered it to be, but I was thrilled to find that the name had stuck and it was still known as Kamikaze Coulee.

We had a new job, fifteen miles south of town, cutting for Chris Pappas. The name Pappas has been shortened over the years from Pappathopalus. Chris was of Greek descent and looked every inch the part.

It was here that we came across the original homestead in which his grandparents had lived. Long abandoned, it proved to be a very interesting place. Following a week's cutting down there, Terry, Charlie and I went to collect our combines one wet day. We had finished and were to gather our machinery and bring it back to town. Due to the weather there was no rush, so we decided to explore.

The original prairie shack stood alongside an old wooden house, built as time went on and as finances improved. The usual collection of dilapidated wooden buildings was in evidence: stables, barns and granaries. Rusting horse-drawn farm equipment littered the yard, the whole place exuding history. Chris turned up and told us what he knew of the place. We looked in the old shack. There was a stove in one corner, an old table in another. All this in the middle of nowhere, several miles from any other dwelling, though I suspect in its day there were others just like it every mile or so.

It was single storey, with a false floor above the kitchen. We asked if anything was up there. 'Doubt it,' said Chris.

Being the lightest I got on someone's shoulder and lifted the trap door. What an Aladdin's cave! There were papers, magazines and documents, all sorts. I related the scene, the others anxious to look. 'Climb up there and pass it down,' said Chris.

We spent the next hour or so going through it. There were magazines called *Successful Farming*, the home-steader's reading of the day dating back to 1917. There were ledgers showing wheat, eggs and other farm produce traded; family photographs, a musical instru-ment like a small guitar, three bowler-type hats called Derbies and other such treasures. Chris was as pleased as we were. He had worked that land since he was a boy and didn't know these items existed.

He generously gave us some of the magazines, and to Terry, Charlie and me a hat each. We have them to this day.

We drove our three 760s back to town, proudly wearing our hats. In the Triple T that night we wore them again, Chris buying drinks for everyone. We had a great evening, the end to a memorable day.

During this time, the local farmer and long-time customer of Dale's, Howard Bailey, now retired, let us know that he had some movie footage of Dale and crew in the 1960s. We were invited to see it one evening. Having worked a long day, tired as we were we did not want to miss this opportunity. Arriving late evening we were given drinks and eats, and sat down to view.

It was at that point Howard told us that he could not find the film in question and proceeded to show us wildlife footage. We tried hard not to show our disappointment and took it in turns to stay awake and ask intelligent questions throughout the film.

In June 2000 when I revisited Gildford, Wade showed me a few minutes' film footage he had of Dale, Larry and crew in the 1960s, and the 510s they were running at that time.

We also had time to visit Hill County fair in Havre this year. Our social life was improving, though never at the expense of cutting. Combining always came first. We would never stop a machine for anything other than the weather, a breakdown or getting to a job where the wheat was not quite ready.

New members of the team at this time included Danny, Tony – another teenager – and an Indian kid called Steve. They were good kids who gave 'Super Trucker' John Fox a hard time.

John had brought his hunting rifle along with him,

leaving it kicking around on the bunker floor. When I suggested he ought to put it somewhere safe he replied that it was okay, there were no bullets in the breach, though maybe the magazine did have some in. I was horrified at this and as he checked it he discovered a bullet *was* in the breach. Damn lucky someone's leg was not blown off!

I had heard that mixed-age crews led to trouble. During the war, custom harvesters hired whoever was available and this was often older guys, too old for National Service. Post-war it became popular to hire high school or college students for the summer. Both age groups had advantages and disadvantages, mature experience against youthful exuberance. But mixed-age crews rarely gelled. We were a predominantly young crew and an older guy like John simply did not fit in. Working and living so close together in that environment sometimes led to a point where you were ready to kill each other anyway, but basically you got on or got out.

The year 1980 was special for Margie because a start was made on her new house at the ranch in Nebraska. She had lived in trailer homes for many years. The footings were dug and a carpenter was hired. He and Margie built the house between them. Made of timber, it was a lovely home.

We really missed not having Margie on harvest and it was only on odd occasions that I witnessed her capabilities. I remember driving from Kansas to the ranch one year to pick up a draper header for the combine. I arrived tired, dirty and hungry. Margie gave me some of Dale's clothes to wear and washed mine while I was having a shower. I had a huge meal of T-bone steak, vegetables and gravy. A cold beer was followed by a comfortable bed in clean sheets. I felt like a king.

The storeman of a Massey dealer in Texas told me that if Margie came in for parts and he could not locate them, she would come round the counter and show him where they were kept! It was bad timing on our part that she should retire from harvest the year we first joined the crew. She was at least half the business, and in total control of the domestic front. So long as the combines were greased, oiled, fuelled and kept at work, Dale was not too bothered if the operator had eaten or not.

The sixty-foot trailer that Dale and Margie used to take on harvest now stayed back at Manchester. From '79 Dale had slept in the 26-foot Red Dale Trailer or sometimes the Terry Trailer. Usually these would be hooked up to water, electricity and had bottled gas. Occasionally there were only hook-ups for the crew's bunker, Dale going in there for a shower after his crew had left for the fields.

With Gildford and Kremlin cut out we headed west to Shelby. Here Charlie and I cut together again for a guy called Leslie Benjamin. Some of this was durum wheat; Leslie had a brother, Delmer, who was an aerobatics pilot, keeping his small plane on the farm. He kindly took us up in it one at a time and, with great relief to me, refrained from looping the loop. From the air the area looked like a patchwork quilt of strips and squares. We camped in Shelby, an oil town, much bigger than Gildford and not nearly as pretty and friendly.

Strip farming was quite common in this area due to its low rainfall, wheat being alternated with summer fallow. The idea was to grow one year's crop on two years' rain. In all the states moisture seemed to be the limiting factor on yield. Most of the land seemed quite sandy and not what heavy-land farmers in England would call good wheat-growing land. Clay soil was often referred to as 'gumbo'.

Summer fallow was usually worked shallow with wide V-type cultivator shovels that cut under the weeds without undue moisture loss from the soil. With few trees or bushes the landscape lacked variety. If you had spent four weeks cutting strips, some of them a mile and a half long, you would have committed murder to cut a square or irregular-shaped field with perhaps a tree or water feature in one corner.

★   ★   ★

Once more my time was up. Phone calls to England told me that the fields back home were clear of both grain and straw and that I would need to get back there to start spreading base fertilisers for the following crops of sugar beet. It was always a tough decision to leave Dale when he was so busy, but he understood my situation. I was leaving a lifestyle and job I loved doing, to return to a different job and lifestyle that also gave me a lot of pleasure.

I reluctantly handed my combine over to Darren Shewster who, with his friend Tim, had been hired a few weeks earlier. As before, Dale took me down to Great Falls and booked me into the Ponderosa Inn. I was becoming a regular customer.

I flew home via Chicago the next day. I knew I would be coming back.

Chapter Five

Wrecked Trucks

OUR third year with Dale saw Charlie, Terry and me running our respective machines from the previous year. We still had Bessie, Loom Bow being traded for a new 850 with charge-cooled engine, giving even more horsepower than our 760s. None of us was tempted to swap.

Dale also hired a brand-new 850 from a Nebraska farmer called Jerome. It came with its own trailer, an A frame on narrow wheelbase which hauled the combine backwards. This 850 had serious problems with hydrostatic oil overheating. I can't remember who the driver was but he had my full sympathy, working in the intense heat for days trying to find out what was wrong.

It turned out to be a union coupling, not blocked as you might expect, but drilled out too small at the factory. I think the problem was finally solved by Massey Ferguson Harvest Brigade technicians. As described in Chapter One, in 1944 the Harvest Brigade title had been coined by Joe Tucker to identify the phenomenon of five hundred Massey-Harris model 21 combine harvesters gathering the harvest in military style. There was a similar program in 1945.

From 1946 onwards the Harvest Brigade title referred to a team of technicians with a parts back-up service that followed the harvest north, basing itself in the Massey dealer's yard nearest to where the main harvest was under way. You might have bought a new machine that year, but it could be two thousand miles from the dealer when it needed warranty work. The Brigade, or the nearest Massey dealer, took care of this. Some dealers, who otherwise would have been overwhelmed by the demand for spares, repairs and technical know-how as harvest moved into their area, were glad of the support they received from the Harvest Brigade.

Dale had hired three brothers and their friend from New Mexico that year. The oldest brother, in his early thirties, was a born-again Christian weighing in at 450 pounds, that is 32 stone or 205 kilos: heavy by anyone's standards. Gerry had been a custom cutter himself, but had gone broke. He saw himself as some kind of foreman, or 'straw boss' as it was known. Terry had again worked for Dale over the winter, and if anybody was to be second in command it should have been him.

Fat Boy, as we called Gerry, though not to his face, would drone on and on at breakfast with bible readings. I wondered if he considered gluttony as a sin. They had rebuilt Mule Skinner, burnt out the previous year, with a new cab and engine. There was a limit to what Gerry could do. Too fat to grease up a combine, he could just about sit in one. Some trucks would accommodate his gut, others wouldn't. The younger brother and his friend were fine to work with, but there was an atmosphere and two distinct camps within the crew.

So it was Burkburnett, Texas; Manchester, Oklahoma, and Leoti, Kansas as usual.

From Leoti we split. Terry and Charlie went to Bennett, Colorado, the rest of us to Harrisburg, Nebraska.

In the early spring of '81 I had received an enquiry from Tim Slessor, the BBC producer who had made 'Yellow Trail', as to the possibility of taking his son Jeremy on harvest with us. Jeremy, at the age of eighteen, had just done his A-level exams and was taking some time out. Although he had no agricultural experience, he had run a tractor at his local golf course. I went to London to meet him and arranged with Dale that he should come over.

Jeremy joined us in Harrisburg and turned out to be a natural truck driver, brilliant at his job, liked by everybody and fitting in just perfectly. I have seen other people not fit in because the work gets to them or they can't stand the lack of regular meals. Then there is picking through dirty laundry for something less dirty than you are wearing or just living on top of a load of guys under stress. On two occasions I have seen grown men reduced to tears by harvest.

Food had been a hit-and-miss affair so far that season. With no cook hired we relied on cafés for our meals. So long as Dale got a good breakfast and had his thermos filled when he was paying the bill, he seemed to be able to exist on coffee and wheat crackers for the rest of the day. I never quite got the hang of this arrangement. By early afternoon I was ready to eat again, and by seven o'clock I was ravenous.

The quality of service in cafés varied from town to town, as you would expect. Without asking, all would provide a glass of iced water free of charge as you sat down at the table, a service you wouldn't get in England. Most offered the usual combinations of eggs, bacon, sausage, ham, hash browns and toast. They would ask how you wanted your eggs: over easy (being lightly fried) through over medium to hard. Scrambled eggs were popular, sausage was more often 'patty' (flat and round) rather than link. You had to specify if you wanted link sausage and sometimes it was not available. Hash browns are grated strips of fried potatoes, usually quite nice but occasionally so greasy as to be unpleasant.

If you ordered tea with no other explanation it would come as iced tea in a glass, the most revolting drink I have ever tasted. Just about everybody except me drank coffee but as a stereotypical Englishman, I ordered a 'hot tea' which came as a tea bag with a mug of hot water. This was fine by me.

Pancakes, served with maple syrup and in quantities such as stack, short stack or long stack were rather disappointing, much thicker and blander than their English counterparts. The waitress would visit the table from time to time to replenish coffee cups – you only paid for the first cup – and to ask if all was well.

Mid-day and evening meals gave you more choice. There was nearly always a Special on the board, say beef or chicken with potatoes and vegetables cooked in large quantities and sold at a discount price. I discovered Lima beans, a bit like our broad bean, and black-eyed peas, as the name suggests, but darker and not as sweet as a green garden pea.

A small bowl of salad was offered with choice of dressing such as Thousand Island, Ranch or Blue Cheese. Hamburger and fries was of course the staple diet. Usually the hamburger was of reasonable quality and meatier than its English counterpart. If you ordered chips you got crisps, so it was always 'fries'. I ate quite a lot of hamburger and fries and enjoyed them immensely. To ring the changes, pork chops or liver and bacon, if available, were my choices.

Combines at Manchester ready to head south to Texas. May 1979.

Cutting in Texas.

Early morning: combines are serviced to prepare for another long day.

The crew, May 1979. Dennis, Rob, Don, John L,
Gary, John C, Terry, Slim (Charlie behind camera).

Ernest and Oscar Goetze.

Loading a header; we are about to move north. Colorado, July 1979.

The convoy is halted by the Highway Patrol as a tornado passes just ahead of us.

Cutting south of Kremlin, Montana, August 1979.

Strip farming in Montana.

Bogged down in Texas; Dale looks on.

Mule Skinner on fire. Oklahoma, June 1980.

Mule Skinner the next morning.

Charlie cuts alongside Kamikaze Coulee.

Loaded and ready to go.

Gildford, Mon

...ical wheat town.

Treasures we found in an old homestead. Chris, Rob, Terry, Charlie.

Leaving the homestead. Montana 1980.

Truck wreck: Bushnell, Nebraska. July 1981.

Jeremy and truck Boris; fortunately nobody was hurt.

Overnight stop en route to Montana. August 1981.

Heading north through Wyoming.

760 combine cutting Montana wheat.

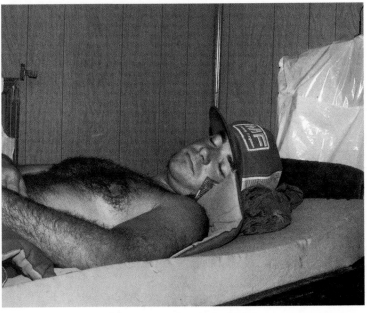

Rob gets some sleep in the bunker.

Oklahoma, June 1982.

Harvesting for Joe Thackston. Colorado, August 1983.

Wendy.

My last road haul. Montana, August 1983.

Rob and Dale.

The crew pay their respects. June 2009.
Kevin, Charlie, Terry, Rob, Steven.

I found myself eating any garnish of cress, lettuce and lemon, peel included, that came my way, to supplement a less than ideal diet. For dessert there would be a choice of pies including chocolate, pecan and cherry. Sometimes we could have ice cream or more exotic things like banana splits. If you asked for jelly you got jam and if you wanted jelly you asked for jello.

By now Charlie and I had slipped into the language and culture very easily and so did not disgrace ourselves when ordering food. Of course we still had our British accents, which often brought friendly chat from anybody who heard us speak. I remember one piece of advice Dale gave me as he was paying the bill to a particularly crabby café owner: 'If you are going to take money off people, you had better smile while you are doing it.'

When cutting we usually ate lunch on the go, and without a cook this tended to be a hit-and-miss affair. We might get the chance to buy sandwich material after breakfast in town. More likely Dale would buy a few loaves of bread and packets of Bologna, a Spam-type meat, and we would stop for a few minutes to grab a sandwich out of the back of his pick-up. Okay when fresh, but three-day-old Bologna out of Dale's pick-up in those temperatures was best avoided. I got into the habit of carrying chocolate-chip cookies so that I had at least something to eat if all else failed; and of course you always had your water jug with you.

The previous harvest I had overheard Dale telling our new cook, Kay, to make sure that Robert got his hot tea for breakfast, which I found rather amusing as he was quite happy for you to starve to death for the rest of the day!

So, back in Harrisburg we were eventually joined by

Terry and Charlie who had tales to tell of combines and girls. With the Mule Skinner rebuilt we had our full complement of trucks. Dale had warned everybody about wrecking trucks. His insurance company was getting jumpy.

In July, north of Bushnell, Nebraska, three loaded grain trucks were on their way to the elevator, climbing a steep hill on a gravel road. The front truck, Dermot, driven by a monkey, missed his gear change, lost his brakes and ran back into Boris driven by Jeremy. It smashed Boris off the road, and ran back into Mo, driven by a Welshman called Richard.

Mo's brakes held both trucks but suffered severe damage to engine and transmission. Boris was a complete write-off. Miraculously no one was hurt, just badly shaken.

We combine drivers some ten miles away heard the two-way radio exchanges and abandoned cutting to go and help. A tractor was borrowed to drag the wrecks off the road. Dale said we had to scoop shovel all the wheat off the wrecked trucks onto other trucks as quick as we could. This task was made easier by a grain auger I commandeered as it was going down the road. Scoop shovelling grain in that heat among all the wrecked trucks was not a pleasant exercise.

Dale was pretty sickened by all of this. I wondered how he stuck the pressure. I could not imagine becoming a custom harvester. Driving for Dale was great, but I didn't want all the crap that went with running an operation that relied on human nature and the weather.

Montana was ready and we were two trucks down. A blue Chevy twin screw turned up, borrowed from a friend of Dale's in the area. We called the new truck

Becky and I hauled my combine to Montana with it. But, with two trucks down, more drastic action was called for. Terry and John Neumeyer in the Jimmy, sharing the driving, hauled a combine as far as they could get in daylight, parked the combine and lowboy and drove back overnight to collect another combine.

Another guy would drive a pick-up pulling a header trailer with them, the header trailer would be dropped off, and the pick-up loaded up into the Jimmy's box for the return journey. There was only enough room for Terry and John in the truck cab, which had bucket seats, so the other guy rode in the pick-up in the truck box. This arrangement carried on for three days until everything got to Montana.

Road trips like this put everybody under pressure, and though the law required a truck pulling an over-width combine to shut down one hour before dusk it seemed like we always ran until dusk and more often than not pulled into town well after sunset. We had a saying: 'Arrived dark-thirty.'

The police could be very strict about traffic offences and Dale had had his share of fines for travelling after dusk. I have to say that any dealings I had with either highway patrol or local police were always good-humoured. I was once pulled over for some minor traffic misdemeanour as I was hauling a combine by truck. When the officer realised I was from England he wanted to know all about my home country and was amazed that my English driving licence was good until I was seventy years old.

His knowledge of Europe seemed somewhat limited for a man with some education. He had a brother-in-law in the military stationed in Germany – and wondered if I had met him! When he looked at my English

99

licence I think he could see the paperwork getting very complicated so I was let off with a friendly word to take more care in future.

Back in Texas in 1979 a pick-up full of crew – three on the seat, six in the back – was stopped driving through Burkburnett somewhat erratically when the driver was trying to avoid a possum on the road. We had shouted to the driver that we had the police behind us but as we were all in high spirits he ignored our warning.

They threw the lights and sirens on and we were pulled over. 'I guess you guys are on harvest?' said one officer.

They scrutinised our identity papers. The Americans all had photo driving licences even in those days, and Charlie and I had photos on our international driving licences. They were checking to see if any of us were wanted men. We might have been fined for careless driving but our status as harvesters enabled the driver to be let off with just a word of warning and instructions to 'Take it easy, boys.'

And so to Gildford where everyone welcomed us. The Starks crew had a reputation, especially in the Triple T.

Levi had bought his own MF 510 combine and was to cut his own wheat. We cut for Wade, Charlie Griffiths, Swede Welsh and others whose names I can't recall. Norman Dack at Kremlin was also an important customer.

As Terry drove Fat Boy the six miles from Kremlin to Norman's farm one morning, he witnessed him consume a full half-gallon of ice cream with a tiny plastic spoon. To quote Terry, 'I sat in awe and fear. I was driving that day, I did not realise someone could wield a plastic spoon that viciously without it melting. I was in fear that Fat Boy would lose one if not two fingers in the process. Yes, half a gallon of ice cream from Kremlin store till you turned left at Norman's, and I didn't drive that slow.'

Fat Boy's sister-in-law would also have made him fried egg sandwiches to eat on his way to the field, despite his having had a huge breakfast. Eventually Fat Boy and family quit. Helping themselves to wrenches from the parts truck and filling their pick-ups and additional fuel tanks with gas charged to Dale's account, they left for New Mexico. They were not missed.

We were camped in Gildford. The bunker was full, and to alleviate the overcrowding I vacated my bunk and moved into Levi and Dian's annexe. One wet day when we were all in town, Levi and family were cleaning their combine down near their house. I went over to help and, having scraped dirt from inside the machine, I dutifully held the hands of the girls who were about two and four years old. We all stood safely back and Levi ran the machine up to eject the dirt.

Now the engine was stopped it was safe to inspect our efforts. Some dirt remained, so I strolled over to our camp to fetch a long piece of metal rod we kept for this very task.

As I returned, Levi was running with Dian in his arms. At first I thought they were fooling around, but then I saw blood, lots of it. Levi put Dian into the back of his car and made to drive away; I looked at Dian, now very grey indeed and we exchanged glances that I will never forget.

It would seem Dian had her hand in the bottom of the unloading auger using the airline as Levi engaged the drive. It was Dian's father who told me this, he was now looking after the girls. Dian's hand was still in the machine and it was our task to retrieve it. I was joined by Charlie, and between us we looked. No sign of it down below so I went up the ladder into the grain tank.

At that point Charlie found the hand and bravely retrieved it. I'm sure if I had found it I would have done what was necessary but, as it was, the situation was now under control, and I quietly went to pieces. I could not bring myself even to look at the hand.

Dave Toner and Jeremy wrapped it in bags of frozen vegetables, and rushed it to the hospital. The hand was relatively intact, but Dian's arm was badly torn up, so stitching it back on was not an option. They were at least able to save her elbow joint.

We were all badly shaken; Dian was like a sister to us. I asked Tom Toner how we would come to terms with all this and he told me that Dian would help us through it. Jeremy thoughtfully arranged for flowers from 'The Starks Crew' to be in her room when she came out of theatre.

Tom was right, she was out of hospital in under two weeks, happy and smiling, a very brave lady indeed. We still had lots of cutting to do in the area, and gradually life got back to near normal.

★ ★ ★

September approached and it was time for me to leave. Dale had an idea. I could return the truck we had borrowed to Chappell, Nebraska on my way home. From Chappell I could catch a train to North Platte, fly from there to Chicago then on to England.

Having said my goodbyes I loaded my kit aboard truck Becky and headed south. Havre, Roundup, Billings, then south-east on 212 to Belle Fourche, South Dakota. Through Rapid City and Hot Springs, I dropped down into Nebraska, through Chadron, Alliance, Sidney and on to Chappell.

This road trip had its moments. As I had no combine with me I could legally travel after dark and had intended doing so. Becky's engine was beginning to miss and splutter and, shortly after dark on mountainous roads in the middle of nowhere, she quit on me. I tried to see what the problem was but it was too dark and turning the engine over continuously was liable to flatten the battery. Getting off the road was the priority.

Walking back down the steep hill I found a pull-off big enough to house us for the night but with an unprotected sheer drop on its edge. Gingerly I eased Becky back. With no engine I had no brake assistance and left her in gear to arrest the motion at intervals. Reversing into the pull-off on the blind side in the dark alone was tricky to say the least, but at last we were just off the road and not in the bottom of the ravine!

I dared not leave the truck's lights on because in the morning I would need all the battery I had. As a heavy truck roared past me up the hill I recognised the need for some kind of marker on the back corner if I was not to be shunted over the edge in the night. Fishing through my luggage I found the very thing, some bright red underwear.

I had a restless night. When I investigated at daylight I found fuel starvation to be the problem. With no tools at hand I had to improvise, the tab off a drinks can serving as a screwdriver.

I needed to get some gas out of the tank and into the carburettor. Those trucks had two gas tanks, saddle and seat. I had been running on the saddle tank, the seat tank still brim full. The previous day I had bought a large

Sprite to go. Ever litter-conscious I still had the card-board cup, its plastic lid and straw in the cab. With the straw dipped into the seat tank I sucked and transferred one straw full of gas at a time into the cup. Then it was poured into the carburettor and, Hey Presto, she struck up and ran.

My instructions were to leave the truck in a parking lot in Chappell, placing the keys in the glove box. I had no contact name, no phone number, nothing. It all seemed a bit secretive, but that was the situation.

I grabbed my luggage, said farewell to Becky and made for the railway station.

Next problem: Chappell is a small town with the traditional railroad running through it, but freight only. No passenger service! Was I going to jump a rail car and become a hobo? I decided against it. North Platte, my destination, lay a hundred miles to the east on one long straight road. I decided to hitch.

It was as hot as hell and I had heavy luggage. Retrieving a piece of cardboard from a trash can, I made myself a sign: ENGLISH – NORTH PLATTE.

Getting a lift was much more difficult than I imagined it would be. I was no longer a harvester, but had become a tourist. For two hours I sat on the edge of town, walking in that heat with luggage not being an option.

Eventually, across the street in my direction came an old red car inside which were two rough-looking, unshaven men. The driver would be in his thirties, his passenger much older, probably in his late sixties. They said Hi and if I would buy them a six pack of beer, they would give me a lift. Unsavoury as they looked, it was my best option. To exchange a six pack of beer for a hundred-mile journey looked like pretty good value for money.

104

Having bought the beer, we set off. Immediately I was on my guard. They said they were on a road trip and started asking questions. I wasn't going to tell them I had just finished harvest and been paid off. I said I had been travelling around, had run out of money, had my air ticket home, and that's where I was heading.

Every so often the car would stop for the old guy to get out and be sick. I couldn't see that drinking the beer was helping him any. They showed me a brand-new watch, obviously stolen from a store, suggesting I buy it for a loved one at home. Again I told them I had no money. After a while they talked of leaving the main road, to find a lake for them to wash. I could see an ambush looming and asked to be dropped off.

They went off the washing idea and we continued. I was taking as much detail in as I could to give to the police later, for I felt that things were going to turn ugly. My plan was to tell them that all I had was a suitcase full of dirty laundry, and that they would have to kill me if they wanted it – rather foolish in retrospect. I figured a kick in the groin would put the old man out of the picture, but the driver was a different story.

In due course we pulled into North Platte. In no part of the journey was I directly threatened, but I was very pleased to get out of that car and among people again. I thanked them for the lift and hailed a taxi to the airport. I have thought about it many times since and feel grateful to have survived the situation unharmed.

Chapter Six

Long Hours

1982 was my fourth harvest. I had intended to do just the one, but it kind of gets in the blood. We had four new machines that year, two 860s and two 850s. We also had the 850 from the previous year and Jerome's 850 leased again. Terry had married Carol from Leoti, Kansas whom he had met during the corn and milo season. Charlie had gone to Australia the previous fall to drive a combine: round the world, round the calendar harvesting. He was still out there but would visit us for a day or so in June on his way home.

We started with a mostly English crew that year, on an exchange program. Some stayed the course, others didn't. Ken and David turned out to be good help. We had had the odd Englishman since 1980, but never this many. It was hard on Terry, grossly outnumbered by Limeys. Jeremy was back, a veteran by now. He had been best man at Terry and Carol's wedding.

Dale hauled the new combines from Canada as before and we did the dealer's pre-delivery inspection ourselves, including assembling the headers and reels. These arrived in kit form and it was like a giant jigsaw to put them together. Terry was spray painting the Mule Skinner truck at base and was left with the crew while Jeremy and I took two combines south to Texas.

It was this year that I had the 'Hitchcock experience'.

If we were going to Texas we usually left the farm and travelled east on dirt roads the seven miles or so into Manchester before going south on blacktop. It had rained hard, the roads were muddy, and Dale said the dirt roads west then south would be better to travel on. Even these were too wet for loaded combines, so we were to road the combines and the unloaded trucks and lowboys separately, about ten miles, to the farm of a guy he knew, just short of the blacktop. We were then to load up ready for the morning and come back to the farm in a pick-up to sleep that night.

It was quite late in the day when we left Dale's farm, armed with a name and directions to our temporary base. I got the impression that Dale had called the guy to ask the favour. By the time we found the place it was quite dark, just starting to rain and a storm was setting in. I was relieved to get everything off the dirt road before it got any wetter.

The place seemed deserted. It had a house on site, and I thought that, as we were expected, somebody might have come out to see us. The name on the mailbox tallied, so I figured we were at the right place. Better knock on the door, I thought.

By now the storm was getting into its stride with heavy thunder and sheet lightning. There was a light on in the hallway, so I knocked on the full-length ornate glass door: no reply. I knocked again, the lightning struck and the thunder rolled.

I got the impression that somebody was there, but there was still no answer. It was now so dark and so wet we could not have gone further even if we were at the wrong place. I persisted with my knocking. A figure appeared behind the glass door. I knocked again, the silhouette stood motionless.

'Is Bill there?' I shouted above the storm.

'He's dead,' came the reply.

At that moment sheet lightning lit up the door and the ghostly figure of a very frightened old woman became apparent. Presumably she got a similar outline of me from her side of the door. It really was like something out of a Hitchcock thriller.

'Dead,' I thought. 'What, just?' It was obvious I was frightening the life out of this poor old lady, but trying to explain the situation through a closed door above the noise of the storm in an English accent was a tall order, and understandably she was not for opening that door. Had he had a heart attack, or had she stabbed him to death with the kitchen knife?

A pick-up drove in at that moment and a guy came over to speak to me. It would seem the old lady had telephoned him as soon as she saw us arrive with the equipment. Apparently Bill had died a couple of years ago; she was on her own and knew nothing of our plan.

I was surprised he took it so well. He said to make full use of the yard, we were very welcome. I thanked him profusely, and asked him to give my apologies to the lady.

When we got back, Dale asked if we had spoken to Bill. I was tempted to say that we had all held hands, closed our eyes and had made some kind of contact with him.

So, Jeremy and I hauled two combines to Texas, left the combines and one of the trucks down there and came back to Oklahoma in the Jimmy for two more. When Jeremy and I made the second trip to Texas we stayed down there on our own. The wheat was not yet ready so we spent several days working on those four brand-new combines. There is quite a lot of work to do on new machines: concave clearance to set, auger settings to adjust, slip clutches to clean of preservative, re-assemble and set up.

We would quit about 6 pm, drive the eight miles to Burkburnett, eat at Sargii's restaurant, then retire to the Twilite motel for a shower. We would eat grapes, drink lemonade and watch evangelists Jim and Tammy Bakker on TV.

Eventually Dale arrived with the rest of the kit and crew and then returned to Terry, still in Manchester. It was my privilege to start harvest that year in Texas, setting up the four brand-new machines, starting them off and then setting the new drivers on their way. These were special times indeed. Jeremy took on our best truck, the Jimmy, and was in charge of the haulage operation.

Harvest started well but turned wet. After heavy rain the ground would become boggy. The intense heat soon dried the crop, but you could be harvesting grain at 12 per cent moisture while plowing ruts a foot deep with your wheels. Usually we pulled the combines out backwards with another combine, but you had to be careful, a twisted chassis was a ruined machine.

One wet year the Goetzes' Steiger Panther tractor had followed us around to haul us out. We emptied grain bins before they got full; it was murder on traction drive belts. Sometimes the water would erode a channel in the soil that was too deep to cross with the combine. These would be cut out either side to avoid damage to the machines.

Dale usually described heavy rainstorms as 'like a cow pissing on a flat rock.' I'm afraid that the crew's description was no less crude: we called them turd floaters.

We were still in the Twilite motel, our breakfasts at

Sargii's restaurant becoming a great social occasion. The previous year we had made friends with another crew working in that area, although normally Dale discouraged fraternisation with other crews because you could lose work that way. But these small operators were no threat to us, they too were parked in town that year and we always ate breakfast with them, Dale included.

We continued to sleep at the Twilite. I am not sure why Dale did not bring the trailers south; maybe he did not have enough drivers. A motel room had two double beds, sleeping four of us, and there were also people sleeping on floors.

Sharing a bed with another male might seem strange, as indeed it does to me now. But back then it was all part of harvest. However, one was a little choosy, and after all Jeremy and I were – and still are – good friends!

In '79, Dale, Slim and Terry had taken one combine from Oklahoma to Texas before Charlie and I arrived. At dark they checked into a motel room for the night, one room with one double bed. Dale said he would sleep in his pick-up and let Terry and Slim share the bed. After a meal Dale went back to the room to take a shower, sat down on the bed and got comfortable. They ended up all sharing the bed, with Terry in the middle.

With Manchester ready we left Terry and Carol to finish Texas. Carol had soon mastered the art of combine driving, and this meant that Terry could truck the wheat. The Oklahoma harvest went well that year. We cut for Larry and other customers in the area.

With four new combines, breakdowns were few and far between. Drivers with mechanical aptitude usually fixed their own problems, but by now Terry had proved himself to be a very competent mechanic and when he was with the rest of the crew would tackle any job that was beyond the capabilities of some of the drivers. Charlie visited us for a couple of days on his way home from Australia where he had been running a Massey 760 combine on a farm in New South Wales.

★   ★   ★

The pressure was on. Leoti was ready before Manchester was finished. We loaded my combine up one afternoon and Jeremy set off with it behind the Jimmy along with another guy towing my header behind a pick-up. They had instructions to get as far as they could that night and phone in at dark.

I stayed back to load the rest of the machines at Manchester when we had finished that night. Then I was to take Larry's pick-up pulling the Red Dale trailer and drive overnight, meet up with the boys en route and get the hell to Leoti. Larry's pick-up had been converted to run on liquefied petroleum gas (LPG). We should have been able to switch this back to gasoline but Larry assured me it would not run on gasoline and calculated a tank full of LPG would not get me there either. There would be no place open to refill with LPG overnight.

Armed with this information I left Manchester some time after midnight and caught up with the boys in Dighton, Kansas. After I woke them up – they were sleeping in the truck box – we set off for Leoti. Whether I pulled into Leoti on fuel fumes only, I don't know, but I made it okay.

We drove to Carl Downes' farm, unloaded the combine, hitched up the header and I started cutting. Jeremy and his helper parked the Red Dale and headed back to Manchester. Sleep? Well no, not till eleven o'clock that night. By that time I had worked forty hours without a break. I was worried that I would be too tired to sleep when the chance came, but I was well past that stage. I climbed into the bed in the Red Dale and went out like a light.

When Carl was cut out I spent a few days cutting for Chuck Baker, an excitable man, rather highly strung. Apparently his wheat was ready and as Dale put it, 'You'd better get up to Chuck Baker's place, he's about ready to jump through his ass.'

Although these farms and their locations were familiar to me by now, new customer locations were quite a different matter. It is hard to convey the vastness of these open plains, with mile upon mile of wheat, dirt roads and damn all else. A typical set of directions to a wheat field might be: seven miles south, three east and it's the section on the north side of the trail. Depending on which state you were in there could be few landmarks to go by.

Back in '79 we pulled onto a section in Montana following just that sort of information. Two-way radio conversation struck up as doubts started to creep in. It does not take long for six combines to cut two tandem loads of wheat, and we were of course on the wrong section. No harm done, you simply move to the correct section and at the elevator book those two loads to their actual owner who gets two loads of wheat cut and hauled for free. I am amazed it did not happen more often.

We were short-crewed and had moved from Manchester to Leoti leaving some of the equipment behind. After a few days cutting we piled several bodies into three pick-ups and drove overnight back to Manchester to collect the rest of the gear. I was riding with Dale, sleeping so I could then relieve him. He had not told me that we were going to stop at a bus station to pick up a new hand, and when we did he told the new hand to drive his pick-up while he slept in one of the others.

I had slept through all of this ('you looked half-ass comfortable and I didn't want to disturb you') so that when I woke I didn't know where I was, where I was going and who the hell was driving me. A broad smile swept across a friendly face and said, 'You must be Rob.'

It was not unusual to collect a new crew member from a bus stop. Maybe they had answered an advertisement in a paper several hundred miles away. Some were drifters by nature, others were in search of adventure, some just in it for money. If they arrived in a car, Dale would always insist they leave it behind when we moved to the next job. Usually they would be needed to drive equipment anyway, but without their own transport on hand they would be less likely to quit, and it would be harder for them to get into town and get drunk.

Dale had bought two second-hand bob (four-wheel) trucks that year. We put a grain body on one and left the other as a flatbed for a second service truck. It was my misfortune to be driving both of them when their gasoline engines blew up.

I was hauling a combine on the return journey from Manchester to Leoti, less than two miles from Carl Downes' farm, when the first truck's engine started making horrible noises, so we towed it the rest of the way. I took the dipstick out; it had chunks of metal on it. The engine was terminal and had to be rebuilt.

The second truck's engine failed dramatically some weeks later when I had been up to the Nebraska ranch to pick up a draper header and was on the run back to Leoti. Broken down, I hitched a lift into the nearest town and tried to call Dale, unsuccessfully, so I called

Margie whom I had left only hours before. She set off in her car and trailer.

The local garage took me back to the truck and towed it into town. I asked to borrow a tractor and loader from a neighbouring farmer to lift the machinery off the dead truck and onto Margie's trailer.

This tractor was in a hay field, about a mile out of town. It was an elderly International with Farmhand loader attached. Having just had one engine seize on me I was keen to check the engine oil level before I started it. The dipstick was proving elusive. I looked everywhere, it wasn't part of the filler cap, in fact it did not exist. Two drain cocks on the sump had to be unscrewed; if oil ran out of the top one all was OK, if it did not run out of the bottom one oil was required. I was on a steep learning curve!

We put the Draper onto Margie's car trailer and she ran me down to Leoti. Dale was under pressure elsewhere and was not amused, but he knew I treated his equipment as though it were my own.

Because Dale tended to use young and inexperienced drivers on the trucks, hand brakes were often disconnected as they were sometimes forgotten and left on. Another favourite was for the PTO to be left in gear after the truck had dumped. This would result in eventual bearing failure.

Two-speed axles were sometimes disabled because you could tear out the axle gears with misuse. We would reconnect them for hauling combines. Humphrey was a straight five-speed with no two-speed axle and for this reason, as well as being gutless, he was my least favourite. Dermot and even Boris had lots more go in them as well as having that magic half-gear. But the king was Jimmy Diesel with that awesome thirteen-speed. Not the easiest of gear shifts till you got the hang of it, but once mastered it was mesmerising.

One year I did a 250-mile trip with Mo pulling combine. The two-speed axle was not working and after about 150 miles the gearbox was making distressing noises in anything but second and fifth gears. I did the last hundred miles in those two gears, which made for an interesting journey. I became quite good at reading the road far ahead to avoid having to stop forward motion completely. It turned out simply to be low oil level and when topped up all was okay with no damage done to the transmission.

We were cutting nineteen miles south of Leoti one night. Terry and Carol were still in Texas, and the rest of the crew had gone to town to eat, leaving Jeremy and me to top up the last truck. At about ten o'clock I dumped my last bin full, parked the combine and climbed into the passenger seat of the Jimmy.

We set off with a very competent Jeremy at the wheel. When it started to rain we still had three miles of dirt road to go before we hit blacktop. Dirt roads when wet are lethal, the surface gets very slick. Combines can only be driven with great care in these conditions, four-wheel drive pick-ups are the best option and loaded grain trucks the least desirable.

If it had been raining when we left the field we would not have attempted the journey, but we had come a couple of miles on dry road and were committed. We got in the middle of the road, one wheel either side of the crown, creeping along in bottom gear, trying to reach blacktop. But the inevitable happened and we slid gracefully into the ditch.

Chit-chat over the two-way prompted us to call the crew who had just eaten in town and were heading for the trailers. Radio reception at that range was poor and the reply we got to our predicament was, 'We can't hear you very well but we think you are on your way, see you in the morning.' Jeremy and I were not best pleased. In those circumstances Charlie or Terry would have driven south until radio reception gave them the full story, and then come to the rescue.

We resigned ourselves to spending the night in the truck, and to giving supper a miss. By now it was raining hard and had turned very cold. It had been an extremely hot day and I was wearing only a tee-shirt and jeans. The door lining on my side was missing and that door was facing straight into the wind. The Jimmy had bucket seats so lying down was not an option. Jeremy had a sweatshirt to put on, and in a gesture of true comrade-ship suggested we wear it in turns, one hour on, one off. There was nothing else in the Jimmy to wear, not even an old sack, the best we could improvise was wrapping paper towels round bare arms.

Tired as we were we must have slept, but I don't remember getting a whole lot of it. After sunrise we climbed stiffly from our seats and lay on top of the tarped wheat, enjoying both the heat and the chance to stretch out full length at last. We must have slept at least half an hour before a plague of black flies drove us cursing back into the cab. It only took a couple of hours of that hot sun for the roads to dry out enough to drive the truck out of the ditch and back into town.

The 850 combine hired from Jerome (second name unknown) had been bought new in 1981 and was hired to Dale for the early part of the season to help offset the cost of ownership. Jerome farmed land up in Nebraska and now needed the machine back for his own harvest.

The previous year John Neumeyer had taken it back.

John, in his forties, was a great character but had a tendency to drink rather a lot. He was gone several days on this trip, ringing Dale to say he was having problems with the Jimmy overheating. Dale said that he had been lied to before. We all knew that John was in the bars, Dale included.

This year it fell to me to return it. The home-made A-frame on which it was towed backwards looked like hell. Indeed it was hell. The wheel-base was narrow and it felt like the whole lot would go over as you went round corners. One tyre was showing wire and there was no spare.

It was to be taken back to the Massey Ferguson dealer who had originally supplied it, Shulte Implement & Co. Butte, Nebraska. Discussing the route with Dale I pointed out that it was against State law to haul a combine in Nebraska on a Sunday and that today happened to be Sunday. Dale said the law didn't count if the combine was hauled backwards!

On all road trips alone such as this Dale would give you a definite route to follow and you stuck to that route. Dale knew where the underpasses were that you could not get under with a loaded combine. You listened as though it was a flak briefing on a bombing mission over enemy territory.

Hauling a combine backwards on the road looked horrible. Several custom harvesters used this method; indeed Terry Pfeiffer had back in '79. When I first saw 'Yellow Trail' I was struck by the beauty of the silhouette of a 760 being hauled – forward facing – behind a tandem truck on one of Dale's lowboys.

Dale gave me a credit card that would enable me to buy fuel en route. It went without saying that I would sleep in the truck and not run up expensive motel bills unless absolutely necessary. Funding road trip expenses sometimes led to interesting situations. One year a port of entry would not accept Dale's credit card and we had to have a cash whip-round off all the crew to proceed. I also recall Jeremy's credit card funding our right of passage for several trucks and combines in 1981. His father, back in London, got the bill a few weeks later.

As I progressed north the tyre showing wire got visibly worse. Luckily they were in pairs and this one was on the inside, so I figured if it blew I could still continue with my journey. About twenty miles from Butte there was a very loud bang and I knew a tyre had gone. Would you believe it! The tyre that resembled a hedgehog was still intact and the good one next to it had exploded.

The wheel-base was now even narrower, and all the weight one side was on a tyre that was showing more wire than rubber. But we made it, and I was greatly relieved to pull into Shulte's yard.

As I was unloading I saw in the row of machinery a 750 that looked vaguely familiar. It was Bessie, three and a half seasons old and looking very tired. The trough underneath the auger in the top of the grain tank had worn completely through, and the auger flights were so thin you could have shaved with them. I had known this combine when it was brand new and was shocked to see her in this condition. She looked like she had cut a million acres.

It occurs to me that I might be giving the wrong impression that the equipment was always breaking down or that it was often raining. If so it is because I am relating events of interest. It must be remembered that we spent weeks and weeks of uneventful happenings when we would get up in the morning, cut wheat all day and go to bed at night. Though I never found this boring, you may think otherwise.

It is easy to think that if you were running the operation yourself you would always have new combines and new trucks but, in the real world, machinery has to be paid for and pay its way. Dale and Margie had not inherited large amounts of money, land or machinery – what they owned had been earned by a hell of a lot of hard work. Plenty of custom cutters before and since have gone broke running equipment that they could not afford.

Dale and the rest of the crew moved to Bennett, Colorado and I returned to Leoti alone to start a new job, harvesting swathed wheat. Unlike a combine harvester that usually cuts, threshes and separates the grain, a swathing machine cuts the whole crop but leaves it un-threshed on the ground in lines across the field. These lines are known as swaths or windrows.

Wheat is usually swathed in the northern states: parts of Montana, the Dakotas and into Canada. Grain ripens unevenly that far north and a week to ten days in the swath allows the crop to mature fully after it has been cut. When it has all ripened off, combines pick up the swathed crop to thresh and separate in the usual fashion.

This Kansas wheat that I was about to harvest had been swathed because it was weedy. What would have been difficult combining was made a whole lot easier by it sitting in swaths for four or five days of hot Kansas sun. The weeds dried like sticks, and the grain sample was much improved. Because the combine did not need to cut the crop the normal cutting table was removed and a gadget known as a pick-up draper header was attached to the combine instead. The technique was to follow in the same direction as the swather to ensure head-first feed.

I had enjoyed picking up those swaths, my first work of this kind in four years because by the time the crew usually hit swathed wheat in September around Cut Bank, Montana and Saskatchewan, Canada, I would have flown home.

When I had finished, I left the draper header at Jim Roland's where it would stay until needed elsewhere. Dale used Jim's yard as though it were his own, and indeed Dale had equipment ranging from useable headers to wrecked trucks strewn over five states. This seemed acceptable to Dale, and after all they have an awful lot of space in the wheat belt.

That year Terry and I took two combines to Montana ahead of Dale and the rest of the crew. Terry and Carol had been stuck down in Texas, trying to finish; continually rained out, bogged down and generally fed up. Terry had told Carol that they would be in Leoti, her home town, by Independence Day (4 July) for sure – we always were. He was so late leaving Texas, Dale sent him straight to Cheyenne Wells, Colorado. They got to spend half a day in Leoti. Carol was not amused.

So, Terry and I were to rendezvous in Bennett, Colorado and get up to Montana as soon as possible. Leaving Leoti and arriving in Bennett I got the job of taking an 850 in for repairs while waiting for Terry to arrive from Cheyenne Wells. One of the new 850s had been consuming engine oil from day one in Texas. The Brigade had suggested that the engine might eventually bed in, but it had continued to consume up to five quarts a day. Back down the trail Terry had helped our college trainee mechanic Clarke to fit a new oil cooler but the problem persisted, so warranty work was called for.

I loaded it up in Bennett and was to take it sixty miles north to Chas M Van Why & Son Inc in Greeley, the nearest MF dealer that did transient warranty work. Dale said to get on the road, he would catch me up and lead me to the dealer's yard when I got near town.

Somewhere north of Bennett I was on some back road about to cross an isolated and very rough looking railroad track. Creeping over in first gear, I got exactly half-way when the wooden barrier came down between the truck and the combine. I was going so slowly I could have stopped with no damage to either combine or barrier, but obviously a train was on its way and I thought it best to keep going. There was a sickening splintering of wood as the decrepit barrier disintegrated, but miraculously absolutely no damage to the combine.

I cleared the crossing and was about to get out and pick up what was left of the barrier on the tracks when a head appeared. Just my bloody luck I thought. The crossing appeared deserted and I had made up my mind to remove the debris and disappear. The head, however, was friendly; it belonged to a track worker, a black man with a broad grin and very white teeth.

'It's okay, man. Keep go'in.' He waved me on. The Gods were with me! It would make a better story to say a huge train thundered by seconds later, but in fact it seemed ages before a very slow, very long freight train lumbered past.

I left the combine, truck and lowboy at Van Why's and rode back to Bennett with Dale. Back in Bennett I met up with Terry and Carol. It was good to see them again; reunions such as this were very special. It had been six weeks since I had last talked with them, it seemed like a lifetime. We swapped stories and set off for Montana.

Normally we would make this epic journey as one crew: there is safety in numbers, and access to the service truck for spares and tools. But Montana was urgent, they wanted six machines yesterday and Dale was still tied up with uncut wheat in Bennett. We set off, Terry in Mo pulling his combine, I in Dermot pulling mine. Carol drove her car and our new recruit, Andy, drove a pick-up pulling the Terry Trailer, Terry and Carol's matrimonial home.

I was feeling ill, that forty-hour shift back in Kansas beginning to catch up with me. We drove on and just north of Billings on Highway 87 stopped for a tyre check. As we pulled away and I changed from first into second gear I lost all drive to the wheels. It was though I was in neutral. The clutch had failed. I felt ill, very ill, like the worst 'flu you have ever had.

I radioed Terry who stopped and came back to help. We decided to unload the combine right there in the road and then tow Dermot out of the way. I went back to the brow of the hill to warn oncoming traffic that the road was blocked. The first car came over the hill at tremendous speed. I'm not sure who was the more surprised, the driver at seeing the road completely blocked, or me seeing the driver engaged in what might be termed 'heavy petting' with a young lady who seemed to be minus several items of clothing. Somehow he missed the combine.

The next car, would you believe, was highway patrol. A stern looking lady got out of the car and to prove she meant business put her hat on. I began to explain the logistics of what my colleagues were trying to do. She looked at me and said, 'Just do what you have to do.' I paused for a moment and with that 'things can't be any worse' feeling, pointed a finger to my head and pulled an imaginary trigger. Fortunately there were female highway patrol officers in Montana with a sense of humour – a broad smile appeared from nowhere.

It was 6 pm. Terry went back into Billings and bought a clutch for a Chevrolet truck. We got Dermot into a field and with the most basic of tools began to remove his transmission.

We took the tarp off Mo and spread it on the ground under Dermot. Rolling the windows down, we placed the length of angle iron we had unbolted from the other truck across from door to door and, with a come-a-long (small hand winch), unbolted and lowered a very heavy transmission away from the engine. It started to rain and hail. My contribution to this operation was limited. Lying underneath that truck I felt so ill that I thought I was going to die. Terry was brilliant.

Carol had stopped off in Billings and was to catch us up. She didn't have a two-way so we could not tell her where we were. This was her first harvest and Terry had told her that we never unloaded combines till we got to the field. She duly caught us up, but we were well off the road with an unloaded combine and she didn't recognise us. After driving several miles she eventually back-tracked and found us. We got so far with the clutch that night when Terry announced we would go back to Billings and get a couple of motel rooms. This I am sure was for my benefit, I was too ill to argue. Normally Terry and Carol would have slept in their trailer, and Andy and I in the trucks.

I was feeling no better the next morning, but I knew I had to get that combine to Montana. Carol could run it once we got there, but neither she nor Andy could haul it. We got Dermot fixed and somehow we got to Kremlin. Unloading at Norman Dack's we were told,

'You're late and you've only got two machines.' Carol took the wheel of Rhiannon, my combine, and ran her like a veteran.

I was taken to Gildford to Tom Toner's trailer where I slept solidly for two whole days. Tom, bless him, looked in at regular intervals and couldn't believe anyone could sleep that long. I had four or five days' rest, Dian Hanson bringing me grape juice and love.

Tom drove me to the hospital at Chester, where they gave me a very large dose of penicillin in the backside.

'Are you allergic to penicillin?'

'No.'

'Are you sure?'

'Yes.'

'Well don't leave the hospital for half an hour because we are going to give you a lot, and if you react he will not get you back in time.'

I didn't leave the hospital for half an hour.

In due course Dale and company caught up with us. Carol had cooked for us since reuniting. We cut for the usual customers, shared tea and cookies with Wade and Edna, and had some pretty wild nights in the Triple T. This bar had always been special to us. It was the one bar where we found a friendly atmosphere of social drinking. Most other bars just seemed like hard-drinking joints. There appeared to be two categories of people: you didn't drink, alcohol was the demon you never touched, or you were an alcoholic with all the negative things that went with it.

Because we camped for several weeks in Gildford most years there was opportunity to visit the Triple T Tavern if for no other reason than to eat a pizza which might be the only source of food at eleven o'clock after a day's cutting. Here at last was a friendly group of people with whom we could drink responsibly. We had parties, dancing and great fun over the five harvests I spent there. Apart from a few isolated incidents where hardened drinkers were not at their best the next day, none of this affected our work.

Dale was occasionally there in the background but never did fully approve though he never discussed it with us. I maintain we were better operators for letting our hair down just once in a while.

★ ★ ★

Once more my time to leave had come and the harvest of 1982 came to an end for me in the Triple T where they threw a leaving party in my honour. Charlene Melby, Lou McCormack, Holly Dees and the rest of the gang organised everything: their generosity touched my soul.

Carol took me down to Great Falls. I knew I was coming back.

Chapter Seven

A Canadian Connection

1983, my final harvest. We started sixty miles further south that year in Olney, Texas. The Goetze brothers had some land there looked after by their brother, Carl. I got the impression it came under the management of Oscar and Ernest. Dale did not usually cut it; maybe that year we had the time to do it before Oscar and Ernest's own wheat was ready.

Due to Dale starting earlier that year, I did not have time to get to Manchester to help with combine preparation. Instead I flew to Dallas, then on to Wichita Falls to be picked up by Oscar and taken back to his house to spend the night. The next day he and Ernest drove me to Highway 44 to rendezvous with Dale and crew who were en route from Manchester to Olney.

It was good to be back. I rode with Dale in his pick-up for the remainder of the journey, and caught up with his news. I knew that Terry and Carol had quit the previous fall to take other directions in life. There was no Charlie and no Jeremy.

The crew that year had something we had longed for . . . a damn good cook! Patty Hogan proved to be a very valuable asset: kind, cheerful, an excellent cook and a competent book-keeper. Her partner, Ken – both in their mid thirties – was a farmer from Iowa who had run into hard times. He had a dry sense of humour and drove the White Freightliner semi truck with hopper-bottom trailer hauling 1,000 bushels (about 25 tonnes) of wheat per load.

Patty and Ken lived and fed the crew in the Mayflower trailer. They stayed the whole season with Dale, quite an achievement, as one year – I think it was 1980 – we got through five cooks in three months. On two occasions we did not even set eyes on the person Dale had hired to feed us. We came home late one night to find burnt offerings in the oven and no sign of a cook. Perhaps she had taken a quick look in the bunker, seen an appalling mess of unmade beds, dirty clothes and girlie magazines, and decided that this was not the place for her.

Up until meeting Patty I had formed a mental profile of the only sort of person – apart from Margie – who I thought could survive the thankless task of cooking for the crew. A forty-year-old divorcee as hard as nails was what I visualised. Dale knew that any young and attractive single female in that environment would cause a whole heap of trouble among his crew. The crew were probably of the opinion that we were prepared to take that risk!

In August '79 Dale called Tim Slessor back in England to ask if he knew of anybody who would come out and cook. Richard Curry turned out to be a great guy but, having driven grain truck in his spare time, found it infinitely more enjoyable than feeding a group of ungrateful, unshaven harvesters. Dale and I were late in one night; Richard had left our food in the oven to keep warm. Dale carved the beef, and between us we ate the whole joint, most of which Richard had planned to use the next day.

Certainly, having one or more females along, be it a

cook or a crew member's wife, altered the dynamics of the crew, and personal hygiene usually improved somewhat among the single guys when we did. Ten single guys living and working together with almost no female contact was an artificial situation. By my fifth harvest I had experienced both scenarios.

In Olney we parked the trailers in the local trailer park. As transient workers we were put at the far end – and more importantly the lowest end – of the park. One night we took seven inches of rain in just a few hours. The nearby creek overflowed and we were flooded out. Water did not quite get into the bunker, but I stepped outside into two feet of flood.

I had a new 860 that year, Stevie. A new hand, Ian, who had joined Dale the previous fall had the other new 860. Ian was useful help and a good person to know. We retained the three 850s from the previous year.

From Olney we did our usual run of Burkburnett, Texas; Manchester, Oklahoma, and Leoti, Kansas. Crew turnover was fairly low so far that season. One year I counted forty crew members in the space of four months, and with an average crew of ten men this equated to four complete crew changes in the time I was there. Some people were quite simply not suited to the job or its life-style. Others would join full of enthusiasm, which you would see slowly drain away.

The job was not well paid, and I think never has been. We got a thousand dollars a month and were supposed to pay for our own food. If we had a cook she would keep account of all the meals you ate as Dale paid her and also the grocery bills. Charlie and I had said we would work for nothing the first year, and so any money was a bonus. In fact I never paid for any of my meals. Dale usually paid my airfare home together with any internal flights. There was little to spend money on because accommodation, be it bunker or motel room, was all taken care of.

There was no 'pay day' as such. Some crew members would have drunk the lot or quit had they regularly received all their pay-cheque. Dale would write out a cheque for a few hundred dollars to some people on a fairly haphazard basis. He would settle up at the end of the season or when somebody quit, though I suspect not everybody got paid in full if they were leaving in less than honourable circumstances.

It suited me not to be paid until the end of the season. I would draw a few hundred dollars from time to time for such things as miscellaneous requirements or perhaps the odd visit to a rodeo. There was nowhere to keep valuables other than in an unlocked suitcase under the bunk bed. Apart from an old coat I had left lying around I never had anything stolen in five years.

Maybe I was lucky. Back in '79 Slim had asked to borrow my watch for the day as he had to get up early for some reason, but I got it back when he had done with it and, drifter though he was, I trusted him. Tom Kirk was less fortunate in 1975: he had his toolbox and his credit card stolen by a fellow crew member. The toolbox and its contents had been given to Tom by his father, so had great sentimental value. The bank was very understanding about the credit card and reimbursed the money that had been fraudulently withdrawn from Tom's account, but Tom never got his toolbox back.

In Leoti we had some hailed-out wheat to cut. Most people in England have witnessed a hailstorm, but on the Plains hailstorms are not only more frequent, they are much more dramatic. Personally I have only witnessed hailstones the size of broad beans, but it is not uncommon to have hailstones the size of golf balls, tennis balls or in exceptional circumstances – grapefruit. Cars left outside in such storms would be severely damaged.

Even hail the size of grape pips could devastate a wheat crop. It shatters the grain out of the head and mangles the straw. A year's work – and income – could be wiped out in minutes. Our hailed-out crop in Leoti was yielding a pathetic four bushels to the acre, only fractionally more than we seed per acre in England. The crop had hail insurance and what we were harvesting was a bonus.

It was difficult to harvest that flattened, tangled straw and impossible to make a clean sample of the grain. The straw broke up and went into the grain bin. The elevator complained and I felt better when Dale returned and was unable to improve on the combine settings.

Elevators did not usually bother us too much; grain moisture content was their greatest concern. Trucks would be weighed and a sample taken to determine bushel weight and moisture content. The upper limit for moisture was usually 13.5 per cent before dockage (deductions) was made, but it did vary from elevator to elevator.

I once had a load rejected at 14.5 per cent. I told the guy that if grain was coming off the combine at that level in England you would jump for joy. He said it was the end of the season and he had nothing to blend it with, so would have to reject it. I hauled it ten miles west to a different elevator where it tested 13.5 per cent and all was well!

Feedlots would take 'wet' grain but of course at a reduced price. We did cut some wheat at 8.5 per cent one year in Texas, dry by anybody's standards.

130

Truck drivers were responsible for keeping count of the loads and the ticket issued for each load. This would be needed to calculate the haulage bill and the overage (yield over twenty bushels per acre), and also to ensure each load was credited to the correct grower at the elevator.

Sometimes land was share farmed and would have to be booked in under different names. For instance we had a customer in Kansas by the name of Mrs Buck. Most of the wheat from her farm was booked in at the elevator as 'Buck', but she share farmed some land with her married daughter so this was booked in as 'Buck & Summers'. The land she share farmed with both her married daughters was booked in as 'Buck, Summers & Williams'.

Dumping at the elevator was a relatively straight-forward job. You would wait in line, untarp the truck and pull onto the scales where you would give your customer's name and have the sample taken. If all was okay you then pulled over the reception grid located within the elevator.

Most trucks had three hatches in their end gate; these would be opened by the guy who had signalled you to stop in the correct place. This was sometimes Mexican labour and a pretty lousy job it was. Hot as hell, not much air and plenty of dust. He would signal when to hoist (tip, I use the American terminology), and when to stop hoisting. He would clean out the corners with what looked like a large metal hoe and close the shutters. His signal meant you could lower the box and go back to the scales.

Conversation with this guy was usually zero, but you made up for it when weighing out. Most of the people were genuinely friendly, and would almost certainly know the people whose wheat you had just delivered. If there was a 'honey' on duty it made your day, and you made the most of it. She probably gave you more time than she gave the poor sod opening end gates.

The elevators would be situated alongside the rail-track if there was one. Some grain might be taken away by truck but most of it left by rail cars. There were huge grain storage facilities in places like Enid, Oklahoma and Hutchinson, Kansas. One of the elevators in Enid had capacity to store 28,280,000 bushels (750,000 tonnes) of wheat. From there it would go onto the domestic market or to the ports to be shipped worldwide.

Sometimes we would haul to the farmer's own bins, usually this wheat was to be saved for next year's seed. This was less popular with the truckers as it deprived them of their trips to town.

Wheat country is also tornado country. The parts of Texas, Oklahoma and Kansas we worked in were particularly vulnerable to such storms – Tornado Alley. Though we were on the edge of a few tornadoes I never experienced the real thing. In 1979 we were in Burkburnett, Texas when a warning siren went off. Six or seven of us went into a communal shelter with several other members of the public and stayed down there for a couple of hours until the all-clear was given.

I had my passport and driving licence in my pocket and, judging by the stories of such storms from people around me, didn't expect to see the rest of my belongings again when we surfaced. But the tornado went round us and the damage caused was several miles away.

The following year we were en route to Montana with the whole convoy – combines and trucks, pick-ups and travel trailers – when a tornado was passing through

Cheyenne, Wyoming. We were stopped on the road by the police and prevented from going any further. Plenty of traffic was coming out of Cheyenne as residents fled for safety. Again we missed the eye of the storm, but it looked pretty grim a mile or so to the north.

The advice if you were caught out was to get into a ditch or culvert and hang on to something substantial. It was no good sheltering under your combine – a tornado would pick it up and take it airborne as though it were an empty matchbox. Most of the farmhouses and even the old homesteads had a cellar or storm shelter built in. Food, water, lanterns and other emergency kit would be stored down there ready. These shelters often had two exits in case one became blocked by debris.

From Kansas Dale took two machines to Bennett, Colorado, then led the rest of us to an area north of Sterling, about thirty miles west of Peetz. Having got us there he returned to Bennett and left us to fend for ourselves.

In that remote part of Colorado, Canadians from a thousand miles north were buying ranch (pasture) land, breaking the sod for the first time in its history and growing wheat. This land was a good deal cheaper to them than land at home and, as they saw it, a good way to expand. They came down to seed it in the spring and again to harvest it in August. Both operations fitted in nicely with their workload at home, which, being further north, came later.

The land had some inherent fertility and was to be cropped fence to fence for a couple of years, then half-crop, half-summer-fallow thereafter. The locals did not welcome this intrusion, die-hard cattle ranchers thinking it sacrilege to tear up what they saw as good grassland. What little I knew of livestock production

133

back home, I remembered one cow to one acre as a rough rule of thumb. I asked what their stocking rate was. One cow to twenty acres!

Joe Thackston and his sons, Brian and Rod, farmed 2,200 acres of this land. They had brought down their John Deere 8820 combine from their farm in Canada, and we were to help them harvest their crop. When working in Colorado, Brian, his wife and three young children lived in the old homestead house; Joe, his wife and unmarried Rod lived in a newly constructed prefab. There were no other buildings of any consequence on the place. Two John Deere tractors were based there, living outside all year round.

This was rugged, isolated country. Though the wheat was growing on relatively flat land, steep outcrops of rock rose up and surrounded it. At night we would lie in our bunks and listen to the coyotes howling in the bluffs around us. Not for the first time did I feel like a pioneer from a hundred years ago.

One place of notoriety was Billy Goat Hill. A grain truck, not ours, had already run out of road that year, overturned and lost its entire load. Although the truck had been recovered, several bushels of wheat remained unsalvaged as a stark reminder to us all. That hill – the road was gravel not blacktop – was so steep that when we finally left the area we ran the combines, trucks and lowboys separately up the hill before loading up to head for Montana.

Patty continued to cook some cracking meals for us. Replenishing food supplies was quite a challenge for her. We were fifty miles north-west of Stirling, thirty of those miles were on rough gravel roads, and even a pick-up would do well to travel at 20 mph over the bad spots. Breakfast would be served in her trailer: bacon, eggs, sausage and toast, and of course my hot tea! I had tried to get the hang of their strong black coffee but still to no avail.

All the wheat fields were pretty close to the homestead where we were parked so she brought our sacked lunches to the field around noon, rather than presenting them to us after breakfast. We had ham, turkey or chicken sandwiches, peaches, pears or apples and a few cookies. Around 7 pm supper arrived: goulash or chilli, sometimes pork chops or steak slices, served with potatoes and vegetables. All beautifully cooked in whatever quantity you could eat, and served with a loving heart.

We had never lived so well! We would stop our machines to eat the evening meal, but in Dale's absence it was up to me to see that this break was as short as possible: those combines would be needed in Montana very soon.

An 850 broke its shaker-shoe shaft and it was my job to fix it. I drove the hundred miles to Rose Brothers, the MF dealers in Lingle, Wyoming, to get a new one, journeying overnight and sleeping in the pick-up to be there when they opened up at 8 am. As part of these repairs I needed a bracket welding. I took it to a local guy about ten miles from the Thackstons' place.

Dave Greener of Greener's Go-Phers ran a welding and repair business with the help of his two sons. This too was a remote homestead but when I expressed amazement that a repair business could survive here I was told they had oilfield work locally. The Greeners were as nice a family as you could wish to meet.

One of the boys, aged about seventeen, was totally deaf and dumb and had been from birth. He had such an open, honest and friendly face that I took an instant liking to him. His communication with his brother was almost telepathic; both knew exactly what the other was saying.

He wanted to show me their trucks and tractors that were scattered round the yard. His brother acted as interpreter, but soon I was doing simple sign language and found it easy to understand what he was communicating to me. He loaded some gravel while I was there and ran the loading shovel like a veteran. Though he could not hear the engine note he could sense exactly when to increase and decrease engine revs.

Dave told me of the time when his son had indicated that there was a fault with the transmission. Neither Dave nor his able son could detect any problem, but a week or two later a bearing seized in the gearbox.

Montana was pushing and Dale was under pressure to get up there. He left Bennett and headed north with two combines. We had three machines. We had another Canadian – friend and neighbour to the Thackstons – also on our customer list; Alan McKay had just a half-section, 320 acres, to do. Because it was weedy Alan had had it swathed and we needed a draper header to pick the crop up. The nearest draper was at the ranch in Springview, Nebraska, 250 miles away.

I fuelled up the Jimmy Diesel and headed north-east. Dale had hired a Quonset for storage near the ranch from a guy called Von Heider but as there was no loader on site I was to go to a nearby farm and see Grafton Hyzenberger who would help me. It was getting late when I got near my destination, so I called on Margie to repeat what I had done a year or two previous: a shower, clean clothes, a meal and a bed. To my disappointment Margie was out, so I drove on the seven miles to Von Heider's place – that too was deserted. So next stop Grafton Hyzenberger. What a name!

This was a real hillbilly place. It was dark now as I knocked on the door, half-expected, but not until tomorrow. I was all for loading the header up that night but the old-timer was having none of it. I was tired and hungry, and said I was going back to Valentine for a motel room. I had tried to sleep in the Jimmy before, but with bucket seats it didn't work. I had Dale's credit card and thought a night's luxury was in order.

'Ain't no sense in you driving sixty-mile round trip,' said the old man. 'You can sleep here.'

Supper was okay, some kind of stew. He said he would show me where I was sleeping, which turned out to be in an outside barn. Opposite was an irrigation engine pumping water that roared flat-out all night, above which I could just about hear hogs grunting and squealing in the next building. Pigs! Where there are pigs there are rats. I just knew I would wake up with one in my sleeping bag. I thought of a comfortable bed at Margie's, I thought of a nice motel room in Valentine, and I tried to sleep on a very hard wooden floor.

Breakfast was bread and molasses. After breakfast Grafton asked if I would mind if he read a little from the Good Book. It was his house, it was okay by me. His wife and hefty daughter were in attendance. He read for about twenty minutes, and then they started on me. Did I follow God's way? Would I like to give myself to Jesus there and then that morning?

I began by being embarrassed, then slightly annoyed. I respected their beliefs but did not like to be cornered like this. Politely I told him I must be on my way, and he gave in. The most ancient of tractor and foreloader was produced and we went to Von Heider's to load up.

Back in Colorado the boys had nearly finished cutting the Thackstons' wheat. Dale called – the phones were very unreliable in that area – and said that I must send two combines north to Montana immediately, I was to stay alone to harvest for Alan McKay. Dale was stressed; he was late getting to Montana and needed all the machines up there.

Ken and Mark, Patty and James loaded two combines, and with two pick-ups and two travel trailers headed north. I went to Alan's seven miles away. I had kept the Jimmy truck. Alan had his nineteen-year-old daughter Wendy with him. Dale had it all worked out on the phone. 'Get that girl to run the combine, and you truck the wheat.' Dale was very good at taking an awful lot for granted!

Alan's half-section had the old wooden homestead still mostly intact. He had chased cattle out of part of it and cleaned it out. It had water and electricity, no telephone and was very basic but cosy. I slept in the cellar on a proper bed, Alan and Wendy in the two bedrooms above. There was no upstairs. Wendy cooked, and life was wonderful.

On the strength of his increased acreage, Alan had that year bought a brand-new 860 combine to replace his MF Super 92 which must by then have been about twenty years old. He hauled the new combine down from his farm in Saskatchewan to the US border to be told he could not enter the USA with the machine as he had no work permit. The fact that he owned land in the USA had apparently no bearing on the subject.

He had no choice but to take it back home, he also had to give assurance to the immigration officer that he would not undertake any work while he was in the USA. Alan was a man of strict principles and took the matter seriously; he had given his word and would keep that word. I think the revised plan was for him to help

Dale in Canada where he could legally operate his new acquisition. Dale had work around Kindersley and North Battleford, Saskatchewan. Joe Thackston must have obtained the appropriate documents, for he had no trouble bringing his combine over the border, he had in fact run his 8820 alongside us in Kansas for a few weeks the previous year.

Ken, Mark and company ran into trouble when, somewhere up near Douglas, Wyoming truck Humphrey quit on them. They were on a main highway reduced to one lane due to road works. Humphrey and combine were blocking the road, Ken could not get in front to tow him out of the way, and the traffic tailed back for something like three miles. It was a nightmare. They eventually got him off the road and into a truck shop to be worked on.

When they tried again the next day they found that the truck would run okay cold, but as soon as they put the combine behind him and he warmed up he quit again. They called Dale. He was having problems of his own and said they must sort it themselves, and get up there immediately. Another day was wasted.

The following day I got a message from Dale via the Thackstons to take my truck up to them. We were to meet half-way in the little town of Morrill, just west of Scottsbluff, Nebraska and, so as not to waste any of my combining time, we were to do this overnight. Brian Thackston came with me in his pick-up to bring me back to Alan's farm. In the early hours, Ken, Mark, Brian and I met up in Morrill. Ken and Mark were still tense and emotional from their ordeal and, as they related their story of a broken-down truck causing a massive traffic jam, they both broke down and cried.

Back at the McKay place I was now without a truck

to haul wheat. The elevator sent a hopper-bottom trailer out and parked it. Sometimes I dumped wheat on the ground to keep the operation going.

The domestic side was idyllic. Wendy would ride with me for hours. The wheat had been swathed with a 14-foot swather, so instead of taking my usual 24-foot swipe I was taking much less. This slowed progress considerably. We could pick up until about 2 am when I would come in and take a bath. Alan would by then have gone to bed and left me a bucket of water with a heating element dropped in it. This one bucket of hot water was supplemented by two cold ones, to give just enough tepid water to bathe in.

Wendy would have fed me earlier, sometimes she would stay till I quit, and we would drink tea together when we got in.

It would be at least 9.30 am before I could start combining that morning and as I only had myself to worry about and one combine to service, I could have slept in till 8 am easily. But at six o'clock Alan would be tramping about on the wooden floor of the kitchen above my head, giving me the impression he thought I should be up. He was, however, very grateful to me for living in such a humble abode. He said his wife back home wouldn't consider living in such primitive conditions. I thought I was in heaven.

It rained and brought harvest to a halt. Brian wanted to take his wife out that evening, and Wendy had volunteered to baby-sit. Alan was out with his pick-up somewhere so, as I was at a loose end, Wendy and I set forth to the Thackstons' to baby-sit seven miles away on the only transport available to us: a Massey Ferguson 860 combine! On our journey home later that night we encountered as much traffic as we had on our way there

– zero. In the seven miles between the homesteads there were only two or three habitations.

At last I was done. Ken came back south 750 miles in the Jimmy truck and I loaded my combine for the last time. And it was a full load. I had a header trailer with me which, with the 24-foot header on, went into the truck box. I had the draper header still attached to the combine. We had to drop the front portion off, there was nowhere to carry it. Seeing no other alternative, I accepted Brian's offer for them to deposit it in Plentywood, Montana on their way home to Canada. This was a long way from Gildford but it was the best I could do.

We said our goodbyes; this had been a very special time for me. I climbed in the cab, found first gear and we were away. I dropped Ken off in Douglas to collect Humphrey. He was fixed. Some ignition wires were breaking down when hot – give me a diesel truck any day!

★   ★   ★

I made the most of that last road trip. Hauling my combine with the Jimmy truck – two machines that I had grown very attached to – up through the rugged country of Wyoming and into my beloved Montana, I reflected over the last five harvests. It had been a dream come true; better than I ever dared hoped for.

I thought of all the letters Charlie and I had written and all the phone calls we had made to get this to happen. I recalled our first meeting with Dale, that first night in the Bunker in Oklahoma with furniture and debris strewn all over the floor, how we had slept so soundly on beds that had seen so many crews which could have told a thousand stories. There was that road

141

trip south to Texas in 1979, a convoy of trucks, combines, pick-ups and travel trailers; the excitement of my first day in the combine seat, and my first road haul pulling combine. I thought of all the people I had met and fields I had cut; of being tired, dirty, and hungry, and the joy of a shower, a meal and a bed.

When I arrived in Gildford I swapped horror stories with everyone. Dale told me that to get himself out of a mess he had hired two old 760s from a machinery dealer and a couple of winos to drive them. The combines kept breaking down, they had no two-way radios for communication, and the winos were useless. They had kept getting rained out in Bennett and finally had to load up and leave with a few acres uncut. Ken and Mark were still mentally scarred by their incident. When they were trying to fix Humphrey with all that traffic behind them, James had just sat in the pick-up and sulked.

It seemed that only I had been enjoying life. Within a few days of arriving in Gildford, phone calls to England told me it was time for me to go home.

Patty and Holly Dees took me to Great Falls. I had said goodbye and thank you to Dale, and again he thanked me for all I had done.

Much as I loved the life, I knew that my future was in England and that I needed to concentrate my efforts on my work back home. My North American harvest days were over.

Chapter Eight

Seventeen Years Later

I kept in touch with Dale by phone and letter over the years and in the spring of 2000 had the opportunity to go out and visit. I flew into Fort Worth, Texas, hired a car and drove up through the wheat belt to see all my old friends and acquaintances.

I was fairly confident of finding my way about once I got into farming country, but my first challenge was to find my way out of Fort Worth and onto Highway 287 for Wichita Falls. That took me an hour and a half!

Eventually I found the road and started to make good progress, too good in fact. I was still getting used to the car and the roads and was finding it difficult to keep below 70 mph, the designated speed limit. Cruise control was fitted but not yet mastered by the driver. I had noticed that everybody seemed to be sticking to the limit despite having powerful cars and a view of the open road for what seemed like miles.

I saw the patrol car on the opposite carriageway, and as I looked in my mirror I could see him crossing the central reservation. I carried on but the second he threw his lights on me I pulled over. Officer Duncan of Clay County turned out to be a very nice guy. I politely explained that I had just landed in the country and was not used to the car or the roads and apologised for my mistake. I think he was a little taken back by my approach. He would usually be called all sorts of names in this situation, but as I saw it I had broken the law (just) and he was only doing his job.

'Appreciate your courtesy, Robert,' he said in a Texan drawl. I was let off with a written warning and nothing to pay. Not a bad start.

My goal was to get to Burkburnett that evening, and check into the Twilite motel if it still existed. I hit town at 9.45 pm and having toured around for a while had to admit defeat and ask for help. The guy said there was a motel in town, but it was pretty rough and full of harvesters, I would be advised to go back to Wichita Falls. So, he had me marked down as a tourist!

The place he described was exactly what I was looking for, and it was still called the Twilite. My smart red Oldsmobile hire car looked a little out of place among all the custom harvesters' service trucks, but I certainly felt at home.

The next day my first stop was the Goetzes. They were expecting me; harvest was under way. Dale had not cut for them since 1992 as he did not run enough machines for their acreage. Yes, Dale was still cutting; he had not retired but had cut down to two machines. The Goetzes now hired three separate cutters, a total of six machines. They said Dale was the best cutter they had ever had, and would still hire him if he wanted work. Dale now started his season a hundred miles further north in Walters, Oklahoma, but always checked that the Goetzes were cut out before he left to go north.

Oscar and Ernest had changed little in the seventeen years since I last saw them. They took me round the farm and were amazed at what I remembered about the place. I spent that night at Oscar's house and bade them a fond farewell the following afternoon.

It took about two hours to get to Walters, ask directions at the elevator for the farm that Dale was working on and locate the place in question.

I knew that Dale was running John Deere combines by then. The Massey Ferguson combine plant in Brantford, Ontario had gone into receivership in March 1988. I saw a John Deere combine and a Ford pick-up in a field, but no drivers. The mailbox told me I had found Glenn Marsh's place. I introduced myself to Glenn, who told me Dale had just finished cutting the last truckload of wheat for him and had hauled it to town.

Half an hour later Dale pulled into the yard. Well, that was a very special moment that I shall not forget. Within ten minutes it was as though I had never been away. We got the machinery into the farmyard and went to town for supper. Dale and his crew of one, eighteen-year-old Jeff, were staying in a cabin, like a low-grade motel room, in town, so after a shower it was the usual custom harvesting sleeping arrangements, Jeff in one double bed and Dale and I in the other.

Next morning it was breakfast at Simple Simon's. Eggs over easy, sausage, bacon, hash browns and toast, with hot tea for the Englishman. We drove out to Glenn's place and started to load equipment. The header trailer, one of those we used twenty years before, had a bad tyre so it was back into town to get it fixed. Ten dollars for a good used replacement and five dollars to fit it. Dale got inspection certificates for the truck and his pick-up. They are supposed to check lights, wipers, tyres etc. The truck was ten miles away and the guy did not even look at the pick-up we were in, he just wrote out two certificates!

We returned to the farm and went inside to settle up with Glenn. Dale had cut 309 acres averaging 28 bushels per acre. Charges were still 12-12-12, this was all figured out on a rough piece of paper, I used my calculator to do the maths. Glenn simply wrote out a cheque and handed it to Dale, no other official document changed hands.

Dale's next job was at Lahoma, just west of Enid, for this he would need his second combine, still back at Manchester. This job was about ten days off being ready. With Dale in his Brigadier truck pulling the combine, Jeff in Dale's pick-up pulling the header trailer and me bringing up the rear in a red Oldsmobile, we hit the road. Travelling north on 44 through familiar places such as Lawton, Apache; Anadarko, Gracemont, and Binger. By 6 pm we got as far as Biscuit Hill, Hinton on interstate 40, and pulled into a motel for the night.

As he had a few days to play with, Dale checked at the local elevator the next morning to see if any cutting was to be had locally. Farmer W Martin lived ten miles east of town. We drove out there and I witnessed the deal being made. Dale introduced himself and me, and said we had a nearly new John Deere combine looking for work. The farmer said we might not want to do his work – I would not have approached it that way – as he had just three hundred acres to cut in six fields and we would have to take the header off to access them.

It was wet after the rain but not boggy ground. He used to have his own MF 510, then a custom cutter owner and son-in-law had done it for the last five years but due to divorce they no longer operated, hence his call to the elevator. Dale said he would be pleased to do the job, but had 1,400 acres to cut in Enid in ten days' time and said he would cut what he could but would have to leave when the next job was ready. The farmer said he would be pleased for Dale to cut what he could, and the wheat would be ready about Tuesday if it did not rain any more. No mention of price was made by either party. That job would take about four dry days and would fill in time quite nicely if the weather held.

As we still had three or four days at our disposal, Dale decided we should leave the combine, truck and header trailer in the motel car park and head back to Manchester. We stopped in Cherokee on the way home to get some supplies and do laundry.

★   ★   ★

It was good to be back at the farm, just how I had remembered it. A new building had been constructed and newer machinery was in evidence. Exploring in the pasture I found several relics of my past: Jimmy Diesel, minus everything rear of the cab; Mule Skinner, cab and engine burnt out for the second time, and Dale's old Chevy pick-up wrecked and robbed. Margie's pick-up was also there, it looked complete but I wasn't sure if it ran or not. I learned that the '74 tandem, Mo, had been refurbished and sold a year or two earlier; the Caddy and Dermot were in Nebraska, condition unknown, and when I journeyed through nearby Waldron I saw Humphrey parked and overgrown with trees.

Dale had told me his hopper-bottom trailer and Kodiak truck were up in Colorado where they had been used last fall on corn harvest. He needed this outfit for the Lahoma job and had formulated a plan: 'How about if we were to go up to Colorado to pick up that Kodiak?' What is a Kodiak? I asked myself. Is it a truck that hauls wheat *and* takes pictures?

Jeff arrived from home and was told: 'Jeff, I guess Robert and I will go up to Colorado to pick up that truck. You go home, spend some time with the girls, and be back here Monday morning ready to be a harvester.' And so another harvest adventure began.

We finally left around 2 pm, Dale driving his pick-up and I riding shotgun as far as Attica. We stopped there to tighten the air-conditioning belt, for which I was truly grateful as the day wore on. I took over the driving, west to Medicine Lodge, then north through Pratt and onto Great Bend. I was so pleased to make this trip, it was a good opportunity to speak with Dale, and he wanted to talk. We talked of Margie and harvest, and of my life back home.

In Great Bend we refuelled the pick-up and the service tank in the back, then Dale drove north to Hays. I took over at Hays and drove heaven knows how many miles west along interstate 70 to Limon, Colorado arriving 9.30 pm (8.30 local time). We checked out the motels. The first one wanted forty-three dollars, so we tried another. They too wanted forty-three, but came down to thirty-eight after negotiation; the Silver Spur was run by two Chinamen.

We ate supper at the Flying J: hot roast beef, potatoes and gravy, with more hot tea for the Englishman. Back to the motel, shower and sleep.

★   ★   ★

We rose at 7.15, and had breakfast at the Flying J, then drove out to Pete Shanks's place, about fifty miles away due west of Last Chance. This was a sparsely populated area; Last Chance looked a fair size on the map, but I could count no more than about a dozen dwellings: no store, no gas station. Pete had put the batteries on charge, but they didn't even turn the engine over. The Kodiak turned out to be a Chevy tractor unit with Cat engine pulling the same hopper-bottom trailer Dale had in the seventies.

The trailer was still in good shape but the truck was 'over the road' and looked like it had done a million miles. It hadn't been run since the previous November and with knackered batteries wasn't going to run now. We tried jumping it from the pick-up but were losing too much current. We drove to Pete's house about a mile away, and fortunately he was in. He came to open his shop to produce a large-capacity charger and tester. One battery was completely useless. Fifteen minutes of heavy-duty charge and the Cat engine roared into life.

We left about 11.30 heading east on Highway 36, stopping in St Francis for a ham sandwich and Sprite out of the cooler, taking care not to stop the truck. The weather was miserably hot; it was okay for me in the pick-up, I had got air conditioning, but the Kodiak hadn't and Dale was suffering. I didn't have a licence to drive a semi, so I could not take over for a while.

We stopped again in Oberlin for a cold Sprite, Dale took a nap in the cool pick-up, he was hot and exhausted. What drove this man on? Turning south at Phillipsburg onto 183, next stop Medicine Lodge. It was about 9.30 and, thank heaven, it had cooled down. We pulled into a car park to check the fuel in the truck. With the aid of a flashlight I could see the bottom of the tank – about a cupful left. I re-fuelled from the service tank in the back of the pick-up, but didn't bother to put any in the pick-up as there was a quarter of a tank left and not far to go.

We headed for Attica, then Waldron. Half-way between these towns Dale veered off the road and nearly lost it. I think he fell asleep, though later he said he just got too far over. We were doing about 55 mph, I thought he'd had it and I was witnessing his demise. It took a while to get my heart out of my mouth and back to where it belonged.

Eventually the blacktop gave way to dirt roads where I needed to keep Dale in sight in case the Waldron turn was not signposted, but he was chucking up clouds of dust and it was difficult, like driving blind through a sandstorm. If he were to stop suddenly I would run into the back of him, but if I let the dust clear he would be way out in front and I would miss the turning. So it was eyeballs on the windscreen and heart back in the mouth.

We finally hit the farm at 10.45 pm. The clock read 991 miles and I was tempted to drive four and a half miles up the road and back to even up the numbers. I was tired, the fuel gauge was on red, but we were home and safe. Dale microwaved chicken and baked potato for me, but he had been eating candy to stay awake so was not hungry. I hadn't cut any wheat, but I'd sure had my taste of harvest that year. *Sleep!*

The following afternoon it was time to say goodbye to Dale, not easy for either of us, but I had more people to see further up the trail. I had one last thing to do before I left Oklahoma. Placing flowers on Margie's grave I wrote the following on the back of my business card:

> *In memory of Margie, a very special lady to us all.*
> *From friends in England, Tim and Jeremy Slessor,*
> *Charlie Norman and Robert White.*

* * *

That was the last time I saw Dale. We spoke often on the telephone and exchanged cards and letters at Christmas. He eventually got used to the time difference and stopped calling me at 2 am to ask what I was doing!

He ran his two John Deere 9510 combines till 2003

then ran a new 9550 for four more years. He would hire a couple of guys to run the combine and truck but would always set the combine up himself and haul it on the road between jobs. In fact after 1989 when he had cut down to three machines he had done much more combine driving than in the days when I had worked for him.

Health issues became more of a problem in those last few years and he passed away in Wichita Hospital on 30 November 2007 at the age of 83. He had cut wheat himself in southern and northern Oklahoma that summer and his machine was in Iowa picking corn that fall. I was pleased he was able to do the job he loved right up to the end.

It seemed very fitting that Charlie's photograph of the six combines in line was on display at the funeral home, was also printed on the back of Dale's eulogy card and would eventually be etched on his gravestone. I know that Dale had become as proud of that picture as we all were when it was taken back in '79.

It was also very fitting that Terry should represent us all at the funeral. I had written a few words that Terry read out at the graveside service:

A Message from England

It has been our privilege and pleasure to know Dale and his family. We remember his love of combines and the harvest: his determination, expertise and humour. Twenty-eight years after meeting him he is still part of our daily lives. We speak of him often. We are proud to call ourselves **Starks' Harvesters.**

Postscript

BY the power of the internet, in 2009 I was able to make contact with one of the original crew members from the harvest of 1975 – the year *Yellow Trail from Texas* was filmed. 'Young Tom' Kirk had been interviewed by Tim Slessor several times during the film, and was one of only two crew members who survived the whole season with Dale.

Tom's father was working for Massey Ferguson and had spent some time with the Harvest Brigade: Twenty-one-year-old Tom was between junior and senior year at college and thought he might like to do a harvest run. Initial contact with Dale was not encouraging. Dale was only hiring married couples that year so that they would be able to feed themselves and enable Margie to concentrate on other work – though it did not work out that way. Tom was hired on the basis that he would take care of his own food.

This is Tom's story:
'We started out on April 1st at the ranch rebuilding the 510s and 750s. Dale bought four new 760s that year. We worked seven days a week from sun-up to sun-down and then a little more, our first day off being 11th July due to being rained out.

'Only one couple had worked for Dale the previous year, the rest of us were new to the job. I made friends with Richard and Tina Resley, and Tina agreed to cook for me. Richard loved the life; he had to have done to have run a combine with no air conditioning for most of the season. Tina hated it but did well to cope looking after a baby and preparing our meals in their small travel trailer. She cooked some wonderful meals for us.

'By the end of the first two weeks half the crew had quit, by the end of the first month nearly all the original crew had been replaced.

'I enjoyed meeting up with the English film crew and liked to hear about life in England. When I ran a rock through the combine Tim came running over to me to ask if I would mind if he filmed Dale and me "discussing" the situation. I figured Dale could not get too mad at me if Tim was filming so I told Tim to stay as close to me as he could. I know Tim enjoyed filming that segment.

'One crew member was called Junior, he was from a military family so everyone was "Sir" or "Ma'am". One morning we were all standing in the yard getting ready to head out to the fields when Junior headed for the out-house to do his duty. After a few minutes Junior came bolting out of the door with his pants round his ankles screaming "Snake! Snake!" It seems a rattler had crawled under the door while he was in there and he wasn't about to stay around to make its acquaintance. Junior didn't stay much after that.

'There are many things I remember about the harvest, funny how some things never leave us. For several years afterwards I would have nightmares thinking I had fallen asleep at the wheel while the machine was still cutting. I would walk to the bedroom door in my sleep, open it only to wake up and realise that I had been dreaming.

'But there were sunsets which I told myself I would never forget, and I haven't.

'There was the shadow of a hawk sitting on the com-bine's cab roof waiting for a rabbit to run out at the end of the row to give him an evening meal. The coyotes walking across the fields keeping a watchful eye on what we were doing. The occasional nights we finished in time to talk with the film crew from England.

'Then there were the mile-long fields when you thought you would never get to the end of the row – and then when you did, having to turn around and go back again. You thought you would never finish cutting that field. And then the times when you had nine combines in the same field and it seemed like we could wipe it clean in a few short passes. There were the personalities of the many people who were there for just a few days and the ones who were there for the duration. I remember finishing a job in Colorado and the family taking us to a nearby lake that evening. We had dinner by the fire and talked and sang all night.

'It was a wonderful time in my life. I had nothing to lose and everything to gain. I stayed on for the duration because I told Dale that I would. He didn't want to hire a single guy but he did so and I felt obliged to live up to my commitment. It gave me the chance to see a lot of wheat fields and a life I didn't want in the long run. But it is something I never regretted doing and added more to my life than I could have ever imagined. It gave me the reason to go back to school to work for a better life for myself.'

★   ★   ★

Thirty-five years after 'Yellow Trail' was filmed the legend lives on. In the summer of 2007 Norfolk farming brothers Steven and Kevin Clarke were travelling through the wheat belt of Oklahoma hoping to find Dale Starks, in what turned out to be Dale's last harvest.

I had met Steven and Kevin quite by accident in 2001 when they came to buy a combine straw chopper from me. They had heard of 'Yellow Trail' but hadn't seen it. They had read all about the North American wheat harvest and had been desperate to make a harvest run, but with no contacts abroad and a busy life in England it never came to be. After that initial meeting, having watched the film, talked for hours with both me and Charlie, and seen Charlie's portfolio of slides they decided they must visit the wheat belt, find Dale and interview him.

So in June '07 they were driving south through Oklahoma when they spotted a Massey Ferguson 750 combine harvesting wheat. That was a rare sight indeed – a 1970s Massey combine at work surrounded by hi-tech John Deeres and Gleaners. They stopped their car and went across to speak to the driver of this combine – one Delbert Joyner. Delbert must have been quite apprehensive when two guys started taking pictures and climbed up the combine steps while he was cutting and then started to ask questions in a strange accent. But of course with the typical way a wheat farmer welcomes strangers – he could sense they were of farming stock immediately – he happily talked about his combine, farm and way of life. 'My wife is bringing supper to the field, do say you'll stay and eat with us.'

And so a chance meeting led to a firm friendship. Having spent a couple of hours with Delbert and his wife, Becky, the boys exchanged email addresses, promised to keep in touch and continued their journey.

That winter the seed of an idea crept into the thoughts of Steven and Kevin Clarke and grew. Delbert had more acres of wheat than he could cut with his one machine so he hired a cutter to harvest the rest. What if they were to buy a Massey combine of 1970s vintage, keep it at Delbert's farm and help him harvest his wheat.

155

Delbert must have had reservations about this idea: to let two complete strangers from another country buy a combine, keep it on his farm and help cut his wheat was a huge leap of faith.

But it happened. Custom cutters Marvin Helland senior and junior, from North Dakota, located, purchased, worked on and delivered to Delbert's farm a 1977 Massey Ferguson 760. This combine was in exceptional order for its age. For the harvest of 2008 it joined Delbert's MF 750 – as well as Delbert's relative from Nebraska, Neil Wheling, and his MF 750 – to recreate scenes that could have been taken straight from the BBC film.

Harvest 2009 at Delbert Joyner's farm was a memorable time for all those involved. For Terry, Charlie and I it was our thirtieth anniversary of joining the crew of Dale Starks. And thirty years on, we were together again, harvesting Oklahoma wheat with our friends on combines just like the ones we ran for Dale all those years ago.

The Clarkes had caught up with Dale in 2007 and interviewed him on film. He was certainly more talkative then than he had been on 'Yellow Trail': maybe he sensed that his time was nearly up. He talked of starting out in 1948, of the people he had cut for and the machines he had run. It brought a lump to my throat when he talked of 'those great years' when Terry, Charlie and I had worked for him.

Steven and Kevin Clarke and Delbert Joyner are the latest chapter in this saga. I hope one day they will tell their story.

★   ★   ★

Many foreign nationals have now made the harvest run. Organisations such as the Ohio State University and

The International Agricultural Exchange Association have made it much easier for British people to experience the North American wheat harvest. They take care of finding prospective employers and obtaining work permits.

The life has changed somewhat over the years. Modern combines have far greater output than the ones we ran. Living accommodation, food and laundry facilities have all improved dramatically, as has health and safety at work. Truck driving tests and safety meetings are now mandatory. It is less harum-scarum than it was thirty years ago, the job is more regulated and has lost a degree of charm in the process.

What we got away with then would not be tolerated today, and I don't know if that is a bad thing or not. Certainly the hours are still long and the experience rewarding: to anybody attracted to try it the once I would say 'Go for it.' Many people have, and they all have a story to tell.

About the Author

Robert White was born and brought up on a small rented farm in Derbyshire.

His father's health and their financial position in 1969 saw him leave the farm and move to Norfolk. Leaving school at the age of fifteen he got a tractor driver's job on a large arable farm and after five years was able to start a modest agricultural contracting business.

With a love of the countryside and operating farm machinery he has had a lot of enjoyment from this simple lifestyle over the years. He has taught combine harvester operation and maintenance to both college students and to more mature operators.

Driving combine harvesters across the Great Plains of North America was a very special time in his life. He remains a close friend to many of the people with whom he worked.

Other Titles from Old Pond Publishing

Yellow Trail from Texas

This 1976 programme from the BBC's 'World about Us' series was the first to show the grandeur of the American grain harvest and it inspired a generation of young men such as author Rob White to join the cutters. It features Dale and Margie Starks, their nine MF combines and the seasonal crews who work their way north with seven months of contract cutting. DVD

Two Thousand Mile Harvest

In one of our most popular programmes Dylan Winter follows ten of the world's largest combines in 1996 on their five-month harvesting journey north from Texas to Alberta. Shows what life is like for the crew of eighteen, co-ordinating moves from site to site and dealing with machines on a grand scale. DVD

Custom Cutters

In 2005 Dylan Winter went back to America to film two combine crews working north through the grain belt. The Fredericks run John Deere 9660STS combines and Kenworth trucks; the Farris brothers operate Case IH 2388 combines and immaculate Peterbilts. The programme includes informative interviews with the drivers, contractors and farmers. DVD

Combine Harvesters Parts 1 & 2

Chris Lockwood's pair of DVDs shows forty-five combines at work on British farms in 2009. Seven models from MF are included – 788, 400, 525, 865, 29, 40RS and 7278 – in a range that starts with a Massey-Harris 21 and concludes with a Claas Lexion 600TT. Or, in terms of header size, from 12 ft to 40 ft. DVDs

Classic Combines: 1930s to 1990s

Beginning when the binder reigned supreme, this archive programme includes rare footage of early harvesting machines, including those manufactured by IH, Ransomes, Fisher Humphries and Claas. The DVD brings the combine story up to the 1990s with extracts from more films by Claas as well as John Deere, New Holland and MF. Compiled by Brian Bell. DVD

Harvest from Sickle to Satellite

A range of cereal harvesting equipment, all working. The sickle and scythe are followed by a McCormick Daisy sail reaper, and a Case model Q. A Lanz self-binder leads on to five combines from a 1944 M-H 21 to a 1998 MF 40. Full script by Brian Bell. DVD

Know Your Combines

In this pocket-sized book for the novice enthusiast Chris Lockwood chooses 43 combine harvesters as examples of those you are most likely to see working on British farms. Each machine is illustrated with a full-page photograph. The accompanying text gives nformation about features and technical specifications. Paperback

Free complete catalogue:

Old Pond Publishing, Dencora Business Centre,
36 White House Road, Ipswich IP1 5LT, United Kingdom
Secure online ordering: **www.oldpond.com**
Phone: 01473 238200   Fax: 01473 238201